T0207203

# SpringerBriefs in Electrical and Computer Engineering

More information about this series at http://www.springer.com/series/10059

Maria de Fátima F. Domingues • Ayman Radwan

# Optical Fiber Sensors
# for IoT and Smart Devices

Maria de Fátima F. Domingues
Instituto de Telecomunicações - Aveiro
Aveiro, Portugal

I3N & Physics Department
University of Aveiro
Aveiro, Portugal

Consejo Superior de Investigaciones
 Científicas - CSIC
Arganda del Rey
Madrid, Spain

Ayman Radwan
Instituto de Telecomunicações - Aveiro
Aveiro, Portugal

ISSN 2191-8112                ISSN 2191-8120    (electronic)
SpringerBriefs in Electrical and Computer Engineering
ISBN 978-3-319-47348-2        ISBN 978-3-319-47349-9    (eBook)
DOI 10.1007/978-3-319-47349-9

Library of Congress Control Number: 2017933566

© The Author(s) 2017
This work is subject to copyright. All rights are reserved by the Publisher, whether the whole or part of the material is concerned, specifically the rights of translation, reprinting, reuse of illustrations, recitation, broadcasting, reproduction on microfilms or in any other physical way, and transmission or information storage and retrieval, electronic adaptation, computer software, or by similar or dissimilar methodology now known or hereafter developed.
The use of general descriptive names, registered names, trademarks, service marks, etc. in this publication does not imply, even in the absence of a specific statement, that such names are exempt from the relevant protective laws and regulations and therefore free for general use.
The publisher, the authors and the editors are safe to assume that the advice and information in this book are believed to be true and accurate at the date of publication. Neither the publisher nor the authors or the editors give a warranty, express or implied, with respect to the material contained herein or for any errors or omissions that may have been made. The publisher remains neutral with regard to jurisdictional claims in published maps and institutional affiliations.

Printed on acid-free paper

This Springer imprint is published by Springer Nature
The registered company is Springer International Publishing AG
The registered company address is: Gewerbestrasse 11, 6330 Cham, Switzerland

# Preface

With the disruptive progress in the field of electronics, ubiquitous wireless networking has become a reality. Smartphones and wireless devices are becoming more of a necessity of our daily routine rather than just a luxurious gadget, as they used to be some years ago. These advances have resulted in a surge in the number of connected smart devices. According to CISCO® report, the number of connected devices has reached more than double the population (2.2x) in 2015 and is predicted to reach more than three times the population (3.4x) by 2020 [1]. The ubiquity of the wireless networking and the abundance of smart devices have led to the invention of the "Internet of Things" concept. The Internet of Things (IoT) refers to the concept of connecting everything, including ourselves and personal items, to sense the environment, communicate together, and come up with informative decisions about their actions, without human intervention. IoT is expected to cover every aspect of our life, starting from waking us up with a fresh brewed coffee and our bath ready, along with a smart car (potentially self-driven) driving us to work, up to every small detail of our life. IoT fields include, but not limited to, smart city, intelligent transport system (ITS), smart housing, eHealth, agriculture, etc., along with new fields that will be introduced in the future [2, 3].

With the emergence of the Internet of Things (IoT) (also sometimes referred to as the Internet of Everything (IoE)), most devices are becoming smarter and connected to one another. Devices are now designed to intelligently act on their own, based on information received from other devices or information about the environment surrounding them. This development requires multiple types of sensors, to be deployed in different environments; hence the sensor industry has become the center of attention of many research efforts in recent years. Due to the huge number of required sensors in the coming generation of IoT, sensing devices need to be cheap, easy to design, and resilient, in addition to their accuracy and instantaneous response. One of the attractive emerging methods is the optical fiber sensing, which has been attracting lots of attention due to its low cost and resilience. The recent reports about coupling optical fiber sensors to smart devices make optical fiber sensing a promising field, with a huge potential market in the area of IoT. Additionally, those reports have increased the growth of optical fiber sensor markets and their field of action. Optical fiber sensing techniques are nowadays widely and commonly implemented for monitoring physical parameters including strain, temperature, refractive index, pressure, humidity, etc., in addition to the monitoring of health parameters (e.g., heartbeat, blood pressure).

Due to a surge in the interest in optical fiber sensing, this brief provides a review of the evolution of optical fiber sensing solutions and its applications. Following such path, a review of the working principles of optical fiber sensing is given. The different production methods are presented and discussed, highlighting their evolution and analyzing their complexity. Under this scope, the brief presents a review of the existing silica optical fiber sensor and polymer optical fiber sensor solutions, comparing its field of action (sensitivity, accuracy), complexity of manufacture, and economical costs. A special attention is given to low-cost production methods. The brief evaluates different existing techniques and assesses the accuracy and suitability of these sensors for possible IoT integration, depending on different considered scenarios.

Our brief represents a timely report of the state of the art of an emerging topic "optical fiber sensing" that is expected to play a big role in the world of the Internet of Things (IoT) and the smart city of tomorrow. The authors believe that this brief should be used as a handbook for readers interested in the field of optical fiber sensing, as well as in the general field of sensors and their production. Readers, new to the field of optical fiber sensing, will find this brief a very useful starting point for their career. Readers are encouraged to read the following outline and guide on how to use this brief, to find the best way to benefit from our work.

## Brief Outline

This book provides a brief overview of optical fiber sensing, with its different sensing techniques, characteristics, and advantages while going through various proposed solutions to overcome their limitations, targeting low-cost production, higher sensitivity, more resiliency, and wider range of applications, by increasing their range of measuring. The book outline goes as follows.

The first chapter is a general overview of the optical fiber sensing concept, highlighting the idea behind using optical fibers for sensing and the different mechanisms inherent to optical fiber light propagation used for measuring external physical parameters. This chapter serves as a perfect starting point to readers, who are new to the field. It introduces the basics and principles of optical fiber sensing, as well as the different methodologies used in measuring external physical parameters.

Optical sensing can be achieved by monitoring the variation of light properties passing through fiber; hence optical fiber sensing can be either intensity-based, phase-based, wavelength-based, or polarization-based. The first chapter describes all those types of sensing techniques, going through the multiple subcategories of each type. The advantages, applications, and limitations of each type are explained.

Additionally, optical fiber sensors (OFSs) are perfect for the manufacture of distributed sensing systems, which are also introduced in this chapter.

After the introduction of basics and principles of optical fiber sensing in Chap. 1, Chap. 2 takes a step forward to discuss the different production methods used for the manufacture of different types of OFSs. Each subsection discusses one type, introducing the different production methods used and the advantages, along with the most suitable applications. The chapter also addresses the challenges within the different methods, along with some proposed solutions to overcome those challenges. The chapter should be treated as a guide into the production methods of optical fiber sensing systems. Readers, already familiar with the topic of optical fiber sensing, would be able to use such chapter as guiding catalogue, which production method to use, based on their preferences; which can be less complex of production method, sensitivity, or cost-effectiveness; etc.

Optical fiber sensing can provide competitive performance compared to more traditional mechanical and electronic sensors, especially when it comes to sensitivity. They even provide more advantageous solutions, due to their compact size and weight and resilience to external harsh environments, along with their immunity to electromagnetic interference. However, for OFSs to reach their potential market share surpassing other traditional types of sensors, OFSs have to provide low-cost production methods. Chapter 3, hence, introduces solutions aiming at designing new and cost-effective production methods for silica-based OFSs. The performance of those cost-effective silica OFSs is compared to similar sensing devices, usually produced by more complex and expensive methods.

All three previous chapters have considered the more common type of fiber used in OFSs, namely, silica fiber. Chapter 4 then introduces polymer (or plastic) optical fiber (POF)-based sensing, which represents an economical alternative solution that has been gaining lots of interest in the research and also on the industrial level. POF offers an attractive cost-effective solution. The various production methods of POF-based sensors are explored. The chapter also discusses the possible physical parameters that can be measured by POF sensors, along with their most common applications. The performance of POF-based sensors is then compared to that of silica OFSs and electronic-based sensing devices.

Having thoroughly discussed the concept of OFSs, exploring its different types and production methods, Chap. 5 concludes the brief by an overview of the application of OFSs in the world of the Internet of Things (IoT). The chapter starts with a brief introduction to IoT and its various application categories. The chapter discusses the technological enablers of IoT and its value proposition. At the heart of IoT, sensing exists as one main and critical enabler. Sensing provides the required context for IoT to exist. Without sensors, smart devices would not be able to acquire the required knowledge about their surroundings, which is a main factor in the smart concept of IoT. Based on this fact, the chapter sheds more light on sensing within the concept of IoT, emphasizing the role of OFSs in such concept. Finally, the chapter presents different applications, where optical fiber-based sensing has been used or has good potential in the future. This list provides the vital link between optical fiber sensing and the future world of the Internet of Things, which acts as the perfect conclusion to the brief.

At the end, conclusion and summary are presented in Chap. 6.

## Reading Guide

This part presents a guide on how to use the presented brief. The brief consists of five chapters, covering different topics, as explained above. Based on the reader's experience and knowledge of optical fiber sensing and the specific question he/she is trying to answer, the reader is advised to read a certain chapter(s).

Firstly, a reader, who is new and unfamiliar with the field of optical fiber sensing, is advised to start from Chap. 1 and go through the brief in the typical order. Chapter 1 provides a good starting point to the field covering the basic concept and principles of optical fiber sensing and the light properties used to sense external parameters, along with the different types and subcategories of OFSs in general.

If the reader is already familiar with the topic, the reader can skip Chap. 1. Depending on the question he/she is trying to answer, the reader can decide which chapter to read. A reader, looking to learn about the production/manufacturing of OFSs, should read Chap. 2. He/she can also find more answers in Chap. 3, with regard to producing lower-cost sensors.

Chapter 3 is also advised for readers familiar with production methods, but who are looking for more cost-effective methods or solutions to overcome the high costs of one certain production method they are interested in.

Readers, who are familiar with silica OFSs and want to learn about other alternatives, are advised to read Chap. 4, which offers information about polymer OFSs, their production methods, applications, efficiency, and limitations.

Finally, Chap. 5 is devoted to readers, who are already familiar with optical fiber sensing technologies, but are new to the topic of the Internet of Things. Chapter 5 provides them with the basics of IoT, listing multiple application domains, where OFSs can fit. It has to be noted here that this brief provides basic information about IoT and how OFSs fit in the world of IoT, but should not be used as a guide of or a book on the concept of IoT.

Although this section guides readers on which chapter(s) to read based on their level of expertise and requirements, the authors still think that for maximum value, this brief is best read as a whole and in the presented sequence.

Aveiro, Portugal                                                    Maria de Fátima F. Domingues
Madrid, Spain
Aveiro, Portugal                                                                        Ayman Radwan

# References

1. CISCO White Paper, "CISCO Visual Network Index: Forecast and Methodology, 2015-2020,"
   June 2016
2. Jankowski, Simona, et al. "The Internet of Things: Making sense of the next mega-
   trend." Goldman Sachs, 2014
3. Al-Fuqaha, Ala, et al. "Internet of things: A survey on enabling technologies, protocols, and
   applications," *IEEE Communications Surveys & Tutorials1,* 7.4 (2015): 2347-2376

# Acknowledgments

The current book is the result of hard work and long hours of the authors. This work would not have been accomplished without the support of many individuals and organizations. The authors would first like to recognize the support and help of all their family and friends, without whom this work would not have been completed.

The authors would also like to thank the Instituto de Telecomunicações, which provided the adequate atmosphere and conditions for performing the required work to produce the book. This work was partially supported by WeHope, an internal project from the Instituto de Telecomunicações, in addition to CELTIC-Plus project "MUSCLES."

This work is funded by FCT/MEC through national funds and when applicable co-funded by FEDER – PT2020 partnership agreement under the projects UID/EEA/50008/2013 and UID/CTM/50025/2013.

Finally, authors Domingues and Radwan also like to acknowledge the financial support of the Fundação para a Ciencia e a Tecnologia (FCT, Portugal): SFRH/BPD/101372/2014 and IF/01393/2015, respectively.

# Contents

# About the Authors

**Maria de Fátima F. Domingues** is a graduate of physics and chemistry at the University of Aveiro, Portugal, in 2005. She received her M.Sc. degree in applied physics and her Ph.D. in physics engineering, from the University of Aveiro, Portugal, in 2008 and 2014, respectively.

Dr. Domingues is currently a researcher at the Instituto de Telecomunicações, Aveiro, holding a national postdoc grant from the Fundação para a Ciencia e a Tecnologia (FCT, Portugal), which is a competitive grant through an open call with less than 15% acceptance rate. She is also associated with I3N-Physics Department, University of Aveiro, Portugal, and the Consejo Superior de Investigaciones Científicas (CSIC), Madrid, Spain.

Her current research interests include new solutions of optical fiber-based sensors and its application in robotic exoskeletons and eHealth.

**Ayman Radwan** is a senior research engineer and EU project coordinator/manager with the Instituto de Telecomunicações, Aveiro, Portugal, working in the areas of 5G and future generations of networking and the Internet of Things (IoT).

Dr. Radwan received his master of applied science (M.A.Sc.) from Carleton University (Ottawa, Canada) and his Ph.D. from Queen's University (Kingston, Canada), in 2003 and 2009, respectively. In January 2010, he joined the Instituto de Telecomunicações (Aveiro, Portugal), as a senior research engineer in the Mobile Systems Research Group within the Wireless Communications Area. Since joining the Instituto de Telecomunicações, Dr. Ayman has been intensively active in European projects. He is currently the coordinator of the CELTIC-Plus project "MUSCLES" and part of the management team of the H2020 ITN SECRET. He previously acted as the technical manager of the FP7 ICT-C2POWER project and the project coordinator of the CELTIC Green-T project. He has participated in numerous events organized by the European Commission, as invited speaker, panelist, and moderator of the panel. He is considered one of the main researchers in Green Communications in the European Research Community.

Dr. Radwan is an active IEEE member for years (currently under evaluation for IEEE Senior Member) and, as a representative of the Instituto de Telecomunicações, is a founding member of the IEEE Communications Society Technical Subcommittee on Green Communications and Computing. He is the author of more than 70 published articles, as well as patents. The focus of his research interests includes next-generation mobile networks (5G), energy-efficient communications, radio resource management, and the Internet of Things (IoT).

# Chapter 1
# Principles of Optical Fiber Sensing

Since the Nobel Prize award in 2009, received by Prof. Charles K. Kao, for his research and pioneering achievements regarding the transmission of light in an optical fiber, optical fiber has been regarded as one of the top inventions of the last decades revolutionizing not only telecommunications systems, but also new fields of applications such as sensing and metrology.

This chapter provides an introduction to the topic of optical fiber sensing. It starts with defining optical fibers in general, giving guide to how to use them in the field of communications, then moving towards the field of optical fiber sensing. The concept of sensing using optical fibers is introduced. The different types of optical fiber sensing are discussed. In principal, different modulation/demodulation principles can be used for sensing multiple external physical parameters. According to those different principles, several techniques emerged for the production of OFSs. The chapter presents a guide for introducing the different modulation types; hence the categorization of optical fiber sensing.

By definition, an optical fiber is a cylindrical waveguide, composed of a core, where the light propagates, a surrounded layer—the cladding with a lower refractive index, and a coating layer for resistance. To confine the light guiding in the core, the refractive index of the core $n_{core}$ has to be greater than that of the cladding to form the total internal reflection at the core cladding boundary, as seen in Fig. 1.1 [1]. Basically, the cladding is responsible for the confinement of the light into the fiber core, decreasing the scattering loss and protecting the fiber core from absorbing the environmental contaminants [2]. The coating of the optical fiber is a layer of material used to improve the mechanical resistance of the fiber and protect it from physical damage and abrasions. The material often used for the coatings is plastic derivate [2].

The light-guiding principal of the optical fiber core is based on the total internal reflection of the optical signal. At any angle of incidence, greater than the critical angle, light will be totally reflected back into the fiber core medium, as shown in Fig. 1.2. The critical angle is defined as the angle at which the total internal reflection occurs, and determined by Snell's Law [2, 3].

© The Author(s) 2017
M.F.F. Domingues, A. Radwan, *Optical Fiber Sensors for IoT and Smart Devices*,
SpringerBriefs in Electrical and Computer Engineering,
DOI 10.1007/978-3-319-47349-9_1

**Fig. 1.1** Optical fiber
basic structure

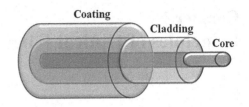

Optical fibers possess many advantages, which include their small size, light weight, low transmission loss, electrical isolation, large bandwidth, security, flexibility, reliability and low cost. The multiple unique features of optical fibers render them suitable not only for telecommunication applications, but additionally other various applications such as laser [4, 5], sensing [6, 7] and medical sciences [8, 9].

The advantages of optical fiber sensors compared to conventional sensing mechanisms are well established; in addition to being smaller and lighter, they provide a large number of measurements with the ability to be multiplexed on a single fiber network.

The general structure of an optical fiber sensor (OFS) system is shown in Fig. 1.3. It generally consists of an optical source, launching the signal into an optical fiber, which will be modulated in the sensing element and detected afterwards by an optical detector (either in a reflection or transmission configuration). An optical detector can be either an interrogator in case of reflective configuration or an optical spectrum analyzer in transmission arrangement, as illustrated in Fig. 1.3.

Modern OFSs owe their development to the laser, and the modern low-loss optical fiber, which led to the first sensing experiments using low-loss optical fiber developed in the early 70s [10]. This innovative work quickly propelled the growth of a number of research groups, focusing on the exploitation of this new technology in sensing and measurements [11–16]. Nowadays, the aim of the research, in the optical fiber sensing area, is to widen the array of fields, where this new technology can be applied. In order to compete with (and eventually overcome) the conventional sensing mechanisms, OFSs must overcome extreme measurement situations and be economically competitive.

The OFS research field has up till now provided sensors able to measure parameters such as strain, temperature, pressure, refractive index (RI), pH, displacement, ultrasounds for bio applications and other sensors, which are used for real time monitoring of structures like buildings, bridges, aircrafts [11–18].

This wide field of applications is possible due to the use of an innovative optical fiber technologies, including fiber gratings, surface Plasmon resonance (SPR), fiber interferometers, Brillouin/Raman scattering, micro-structured fibers, etc. Also, the progress in the research of human-friendly smart materials incorporating fiber devices, including health monitoring systems (i.e. e-Health), is one of the most promising future technologies. Nevertheless, despite its wide application and the growth of research in OFS, its commercialization level is not totally established yet, partially due to high costs of some optical devices [9].

**Fig. 1.2**  Representation of total internal reflection inside an optical fiber

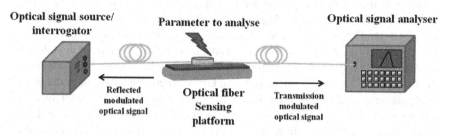

**Fig. 1.3**  Basic components of an Optical Fiber Sensor (OFS) system

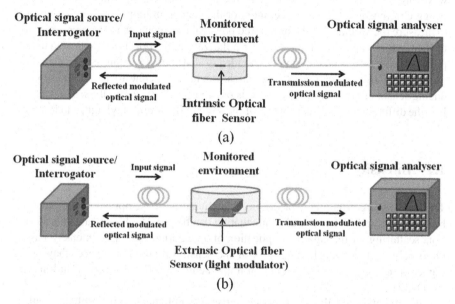

**Fig. 1.4**  Schematics of (**a**) Intrinsic and (**b**) Extrinsic OFS types of optical fiber sensors

Optical fiber sensors can simply be categorized as either intrinsic devices, where the sensing mechanism operates within an element of the optical fiber itself, or extrinsic devices where the optical fiber is used only as the information carrier, to couple the optical signal to and from the region to be monitored. In extrinsic devices, although the modulation is external to the fiber, it can be mechanically attached, and coupled/decoupled [10]. Figure 1.4 presents the basic concept of intrinsic and extrinsic optical sensors.

Another possible categorization of OFS is based on the type of modulation/demodulation process and operating principle. In such categorization, the OFS can be classified as phase, frequency, intensity, or polarization sensor. Due to external perturbations, these parameters values change and by detecting its shift/variation, the external perturbations can be sensed/measured [2].

Different modulation/demodulation principles can be used for the sensors production. According to those principles, several techniques emerged for the production of OFSs. The current chapter introduces the different techniques, which are most commonly used. Each section of this chapter explores one type of modulation/demodulation (i.e. type of OFS) and its most commonly known sub-categories/type. The first section introduces intensity-based OFSs, where the variation in the monitored parameter is sensed by the variation in the intensity of the received light at the optical detector. Section 1.2 discusses the concept of phase interferometer, and lists the different categories falling under such concept, including Mach Zehender, Michelsen, Fabry Perot Interferometer, and Sagnac interferometer. In Sect. 1.3, wavelength based OFS is presented. Wavelength-based OFSs, which include Fiber Bragg Gratings and Fluorescence, monitor the change in light wavelength in order to measure the variation in the monitored physical parameter. Another characteristic of light, which can be used in sensing, is polarization. Section 1.4 is dedicated to the polarization-based sensors. Finally, the idea of distributed sensing is introduced in Sect. 1.5. Distributed characteristic, which provides an economical method of producing sensors to monitor larger areas, is one more advantage of OFSs. The section lists the different types of distributed OFSs, namely Brillouin, Rayleigh, and Raman.

## 1.1  Intensity

The first sensing principle is based on the attenuation of the optical signal induced by the monitored parameter. This attenuation can be provoked by means of absorption, scattering, or through the application of an external force [3]. Intensity-based OFSs rely on the optical signal loss of the sensing modulator; hence, they need higher optical signals intensities often implemented through the use of multimode fibers [17].

The advantages of these sensors comprise its implementation simplicity, multiplexing features, low cost, and ability for distributed sensing. Nevertheless, they require the use of a referencing system in order to mitigate the false readings, due to relative measurements and variations in the intensity of the light source [18].

To produce/induce a change in the optical signal intensity propagated in an optical fiber, different mechanisms such as microbending, evanescent field and fiber coupling can be used [17–19], which are further explained below.

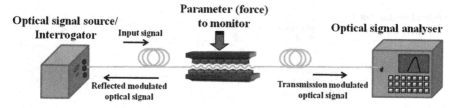

**Fig. 1.5** Microbending sensing mechanism

## *1.1.1   Microbending*

The theory behind Microbending sensors is based on the fact that mechanical periodic micro bends on the optical fiber can induce attenuation of the transmitted optical signal; hence the signal propagated in guided modes can be radiated into the non-guided modes. As depicted in Fig. 1.5, the sensing mechanism can be formed by two grooved plates, with an optical fiber in-between them. When the bend radius of the fiber changes (due to the grooves movements) and surpasses the critical angle necessary to enclose the optical signal in the optical fiber core, the optical signal is attenuated, and therefore the intensity modulation of the signal can be obtained [18].

   This technique relies on two main phenomena: the radiation of the optical signal to non-guided modes of the optical fiber due to the small radius of curvature induced in the fiber; the second one is the resonance effect created by the periodicity of the microbends, which enables the coupling of two modes.

   Microbend OFSs have been explored to monitor a large number of physical and chemical parameters such as strain, pressure, displacement, temperature, vibration, humidity, gas detection, pH, and ion concentration, with acceptable sensitivity levels [17, 18]. When compared with other OFSs, these types of sensors are highly advantageous, due to their design simplicity and low cost fabrication; therefore being suitable for both laboratory and home applications.

## *1.1.2   Evanescent Field*

Another popular intensity-based OFS is the evanescent field based sensor (see Fig. 1.6). Evanescent field sensors use the optical signal irradiated from the fiber core into the cladding, to detect and monitor a specific parameter. This technique is typically used in the fabrication of chemical sensors. For its formation, a section of the fiber (sensing section) is stripped (the cladding can be removed or reduced by polishing or by chemical attacks) and then a light source with an adequate wavelength is used for the detection of the chemical parameter under analysis. The change in optical signal intensity (due to its interaction with the environment monitored) indicates the presence of a specific chemical element and its concentration.

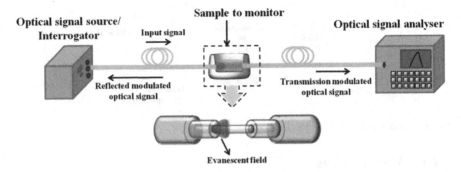

**Fig. 1.6** An evanescent field based sensor working principle

Other versions of this type of sensors are produced with a similar technique, having the cladding replaced by a coating material, whose optical properties can be modulated by the chemical medium in analysis [19–22].

The evanescent field phenomenon is observed, when the total internal reflection in the optical fiber core is not met. Although typically, the optical signal is totally reflected, a portion of it can still escape the interface and penetrates the medium with a lower refractive index, creating an evanescent field [21].

To achieve the basic principle of the evanescent field, the sensor needs to be in contact with the substrate in analysis, so the optical fiber core can achieve a reliable sensitivity level. The presence/absence/concentration of the parameter under monitoring can then be detected by the absorption of the evanescent field wave and consequent optical signal modulation.

The main issue of this type of sensors is related to the potential poor interaction between the substrate and the evanescent field, which is proportional to the cladding penetration depth. The increase of this interaction can be achieved through the use of special fibers (D-fibers or micro-structured fibers) or tapered fibers [21, 22].

### 1.1.3   Two Fiber Coupling

Two fiber coupling sensors are based on a configuration, in which the optical signal from an optical fiber interacts with an external substrate under analysis, and the reflected or transmitted signal on such interaction is coupled in a different or the same optical fiber. This type of sensors can be introduced in different structures/configurations and using different type of fibers (mono or multi-mode fibers). The two fibers coupling can be in a reflection or transmission configuration, as shown in the two parts of Fig. 1.7.

The key element in such sensors relies in a perfect coupling between the two fibers in use. Once the optimum coupling is achieved, this type of sensors is suitable to monitor parameters such as pressure, displacement, positioning, and vibration among others [23].

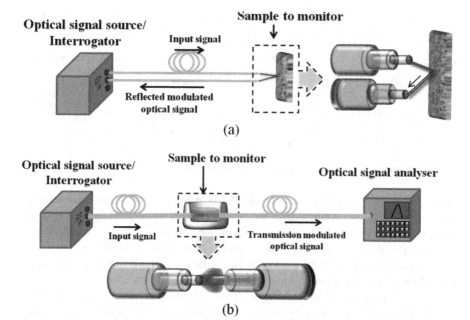

**Fig. 1.7** Schematics of two fiber coupling sensors configurations: (**a**) Reflection and (**b**) Transmission

## 1.2 Phase Interferometer

Interferometric optical sensors are known to provide high sensitivity and accurate results. However, for such results to be achieved, this technique must be applied with precision for the monitoring of specific physical parameters. Also, the use of this technique requires an extra care, since cross-sensitivity problems can often appear, if the isolation of the devices is not completely achieved from the influence of parameters other than the monitored one, such as environmental temperature [11].

Although several configurations can be exploited and will be explored in the following sections, a basic optical fiber interferometer uses the interference between two signals for the signal modulation. This interference is achieved due to the different optical paths the signals travel, which can be made in one single fiber or through the use of two optical fibers [11, 24, 25]. Definitely, it is important that one of the optical paths is exposed to the external perturbations, so the signal can be modulated according to the monitored parameter. The signal modulation can be based in the shift of the wavelength, phase, intensity, frequency or bandwidth of the spectral/temporal information.

With these sensing capabilities, interferometric sensors can provide remarkable performance with high accuracy and sensitivity with an extended dynamic range. The challenge, regarding this sensing mechanism, is size reduction of the sensors

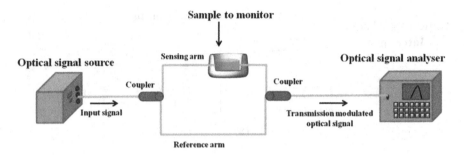

**Fig. 1.8** Schematic diagram of an MZI sensing configuration

without compromising their accuracy and/or sensitivity, to be suitable for micro-scale applications. Such miniaturization requires the replacement of the bulk optical equipment with smaller fiber devices [11]. In-line structures, characterized by having two optical paths in one single fiber, are the best candidate for the implementation of tiny optical fiber interferometers. This solution has high coupling efficiency, easy alignment, and high stability; hence have been widely investigated [11].

Optical fiber interferometers can be classified into four main groups, namely Mach-Zehnder, Michelson, Fabry-Perot, and Sagnac. The four types are discussed in separate subsections below, in which their operating principles and potential applications will be explored.

### 1.2.1  Mach Zehender

Mach-Zehnder Interferometers (MZI) are flexible in their configurations and therefore suitable for a wide number of sensing applications. The basic configuration consists of two arms (optical fibers), independent of each other, which are the reference arm and the sensing arm, as shown in Fig. 1.8.

The optical signal emitted by the optical source, is split (by an optical fiber coupler) into the two arms and then coupled again into a single fiber and onto the optical signal analyzer (OSA). The signal reaching the OSA is modulated according to the Optical Path Difference (OPD) between the two fiber arms. As the reference arm is not exposed to the external/environmental variations, the shift in the interferometric signal is only related to the sensing arm signal. These variations can be induced in the sensing arm by parameters such as strain, refractive index (RI) or temperature and can be easily accessed through the analysis of interference signal shift.

In contrast to the use of two arms (two fibers) for MZIs production, new solutions using only one optical fiber have been reported [25–29]. In this in-line type of MZI, both arms (reference and sensing) have the same physical lengths with an OPD, but the cladding mode beam has a lower effective index than the core mode, which induces a modal dispersion of the optical signal.

**Fig. 1.9** Configurations of
in-line waveguide MZIs
using (**a**) two LPGs, (**b**)
optical fiber core
mismatch, (**c**) air-hole
collapsing on a crystal
fiber (**d**) Small core SMF
fiber

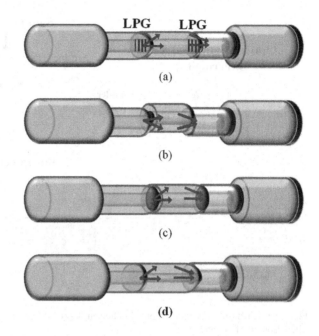

This innovative scheme of in-line waveguide interferometers can be accomplished using different techniques, as presented in Fig. 1.9.

In the configuration shown in Fig. 1.9a, a part of the optical signal in the guided modes in the core of a Single Mode Fiber (SMF) is coupled into the non-guided modes in the cladding by using a Long Period Grating (LPG). The signal in the non-guided modes is then re-coupled again into the core guided modes using also an LPG [25–27].

In the configuration represented by Fig. 1.9b, the beam splitting is obtained by splicing two fibers with a lateral offset and creating a mismatch in the fibers core, which enables the possibility of part of the signal to be transferred from the core modes into the cladding modes [11, 28].

Another technique used for the production of in-line MZIs is based on the air hole collapsing in crystal fibers, depicted in Fig. 1.9c. In this configuration, the core mode optical signal is diffracted in the air-hole and part of it is coupled into cladding modes [11, 28].

Lastly, Fig. 1.9d shows the configuration of an in-line MZI, which uses fibers with different core diameters to obtain the OPD needed. For the production of this MZI, a fiber with a smaller core is spliced to a normal SMF. In the fiber with a reduced core, the optical signal is guided not only in the core modes, but also in the cladding modes [29].

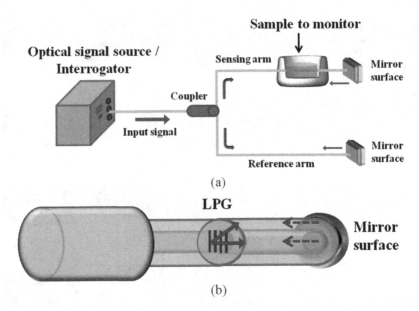

**Fig. 1.10** (**a**) Basic configuration of a Michelson interferometer and (**b**) schematic of an in-line Michelson interferometer

## 1.2.2   Michelsen

Michelson interferometers (MIs) also need two arms to generate an interference signal, but in this particular configuration, each optical signal is reflected at the end tip of each arm, and coupled again into the OSA, as represented by the schematics in Fig. 1.10a [30, 31].

In a vague approach, one can see an MI configuration as half of an MZI, with the existence of a mirror surface representing the main difference between them. Although their production methods and operating principles are similar, the fact that MIs use reflection modes makes it possible for them to become compact and practical for use and installation. Moreover, the possibility of connecting several sensors in parallel also provides them with multiplexing abilities [11].

Similar to MZI, it is also possible to create an in-line configuration as shown in Fig. 1.10b for MI type of sensors [30]. In this configuration, the core mode optical signal is coupled into the cladding mode(s). Both signals, the one propagating in the core mode and in the cladding modes, are reflected by a mirror surface at the fiber end tip. Several optical fiber sensors, based on in-line MIs, have been proposed for monitoring of parameters, such us refractive index and temperature [30–34].

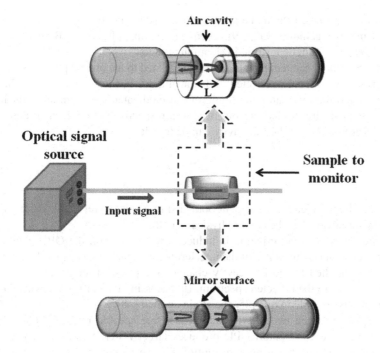

**Fig. 1.11** *Top*: Extrinsic FPI sensor, in which the air cavity is formed by a supporting structure; *Bottom*: Intrinsic FPI sensor, where the cavity is formed within the optical fiber

## 1.2.3   Fabry Perot Interferometer

A Fabry-Perot interferometer (FPI) is generally formed by two parallel mirror surfaces with given distance ($L$) between them. The interference happens due to the interaction between the reflected and transmitted optical signal at the two mirror surfaces. Regarding the optical fiber FPI, it can be produced by creating reflecting surfaces outside or inside the optical fiber [35]. Considering the place, where the reflecting surfaces are built and implicitly where the interferometric cavity is created, the FPI sensors can be classified as extrinsic or intrinsic FPIs. When the interference is created by an external parameter formed outside of the optical fiber, the resulting FPI is extrinsic, as shown in the top part of Fig. 1.11. On the other hand, intrinsic FPIs have the reflecting components (cavity) within the fiber itself, as the bottom part of Fig. 1.11 shows [36, 37].

The main advantage of the extrinsic cavity lies in the possibility to obtain a high finesse interference signal, since it can use high reflecting mirrors. Additionally, its production is quite simple and can be cost effective, as it does not require expensive devices. Nonetheless, the extrinsic cavity FPI sensors have disadvantages regarding the coupling efficiency and accurate alignment [11].

On the other hand, in the case of intrinsic FPIs, the coupling efficiency is not a problem as the reflecting components are within the optical fiber. The interference cavity of the intrinsic FPI can be produced by several and different methodologies such as micro machining [38, 39], chemical etching [40], fiber Bragg gratings (FBGs) [41], thin film deposition [42], among others.

The spectrum generated in an FPI cavity is related to the optical phase difference between the two signals, reflected and transmitted. For a given wavelength, the maximum and the minimum amplitude of the modulated spectrum are indication that both signals are in-phase or out-of-phase, respectively, in a $2\pi$ modulus. The phase difference, $\delta_{FPI}$ of the FPI cavity can be translated as [11]:

$$\delta_{FPI} = \frac{4\pi}{\lambda} nL \qquad (1.1)$$

where $\lambda$ is the wavelength of incident optical signal, $n$ is the refractive index of the cavity material, and $L$ is the cavity physical length.

Sensing wise, if a perturbation is induced into the sensor, the OPD generated influences the phase difference of the interferometer signal. As an example, longitudinal strain applied to the FPI cavity changes the physical length of the cavity, inducing a shift in the reflected spectrum. By accessing the shift in the wavelength spectrum, it is possible to quantify the strain applied [38].

The space between two interference peaks, free spectral range (FSR), is also influenced by the OPD (short OPD produces larger FSR). A large FSR means a wide dynamic range, but a poor resolution for the sensor, and therefore, in the design of the sensor cavity, it is important to balance the ratio between the required dynamic range and the resolution obtained with it [11].

Also, changes in the RI of the cavity material result in phase variation of the reflected spectrum [43]. For RI monitoring, extrinsic FPI sensors are more suitable, since samples can access the cavity [44]. Although the use of intrinsic cavities has been reported for RI sensing, the design of the sensor generally requires elaborate and expensive laser machining process [39]. Nevertheless, a low cost fabrication process has been recently reported [12, 38, 44, 46].

In addition to extrinsic FPI sensors, a variety of intrinsic FPI sensors have been developed with various fiber structures [45]. Among them, a double cavity structure fiber FPI is unique and interesting [35] and a new low cost high temperature sensor has also been proposed [46].

## 1.2.4   Sagnac Interferometer

Sagnac interferometers (SIs) are easy to fabricate, presenting a simple and robust structure. Such advantages brought them to the spot light; hence they became of great interest for several sensing applications [47, 48].

An optical fiber Sagnac Interferometer is basically composed of an optical fiber loop in which two optical signals, with different polarization momentum, are guided in counter directions as illustrated in Fig. 1.12. The input optical signal is split into

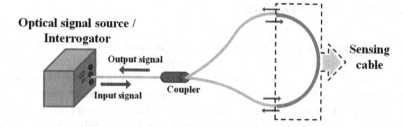

**Fig. 1.12**  Schematic of a Sagnac interferometer configuration

two directions through the fiber coupler and the two counter-propagating beams are re-coupled again. The reflected signal, with the interference spectrum, is then detected by the OSA/Interrogator system [47].

In SI sensors, the OPD is determined by the speed of the guided mode along the loop, which is polarization dependent. In order to maximize this polarization-dependence effect, birefringent optical fibers are often used for the sensing arm. The output signal is a result of the interference between the signals polarized along the fast and the slow axis. The $\delta$ interference phase can be translated as follows

$$\delta_{SI} = \frac{2\pi}{\lambda} BL, B = \left| n_f - n_s \right| \tag{1.2}$$

where $L$ is the length of the sensing fiber arm, $B$ is the birefringent coefficient, and $n_f$ and $n_s$ are the effective indices of the fast and slow modes, respectively [11].

These sensors can be effectively used for temperature sensing (doped fiber has high thermal expansion coefficient inducing a high birefringent oscillation) [49, 50]. For monitoring of parameters such as pressure, strain or torsion, the use of polarization maintaining fibers may be required; hence high birefringent characteristics can reduce the sensing capacity due to their strong temperature dependency [50].

## 1.3   Wavelength

The sensing devices in this category are based on the principle that any variation in the monitored parameter will induce a shift in the wavelength of optical signal. Bragg Grating based sensors and fluorescence sensors are the two most famous examples of wavelength modulated sensors, which are introduced in detail in the following two subsections.

## 1.3.1  Fiber Bragg Gratings

Fiber Bragg gratings are sensing elements, which are inscribed by ultraviolet (UV) lasers into the core of an optical fiber. In addition to all the advantages of an OFS, these devices have an inherent self-referencing ability and can be easily multiplexed along a single fiber. Its field of action is wide, from structural health monitoring to smart materials incorporation, and can even be introduced in bio applications [51].

The Fiber Bragg Grating (FBG) sensor system operation principle lies in the monitoring of the wavelength shift of the optical signal reflected by the grating, which is called Bragg wavelength. Bragg wavelength, $\lambda_{Bragg}$ (wavelength of the signal reflected), which is a function of the monitored parameter (strain, temperature, etc.), is correlated to the effective refractive index of the fiber core, $n_{eff}$, and the grating period, $\Lambda$ by [51]:

$$\lambda_{Bragg} = 2n_{eff}\Lambda \qquad (1.3)$$

From Eq. (1.3), it is clear that the Bragg wavelength is influenced by changes in the grating period (used for strain sensors) or the effective refractive index (used for temperature sensors) [52]. The relation between the strain and temperature variation and the Bragg wavelength is given by the following equation, where the first term represents the strain effect on $\lambda_{Bragg}$ and the second term describes the temperature effect [52]:

$$\Delta\lambda_{Bagg} = \lambda_{Bragg}\left(1-\rho\alpha\right)\Delta\varepsilon + \lambda_{Bragg}\left(\alpha+\xi\right)\Delta T \qquad (1.4)$$

where $\Delta\lambda_{Bragg}$ represents the change in the Bragg wavelength, $\rho$, $\alpha$ and $\xi$ are the photoelastic, thermal expansion and thermo-optic coefficients of the fiber, respectively; $\Delta\varepsilon$ is the change in strain and $\Delta T$ is the change in temperature [52].

An FBG sensing system configuration is schematized in Fig. 1.13. A spectral broadband optical signal emitted by the optical source (interrogator) is launched into the fiber. The grating reflects a narrow spectral component of the emitted signal at the Bragg wavelength. In the transmission analysis, the Bragg wavelength component is missing from the observed spectrum. The details of such configuration are illustrated in Fig. 1.13.

FBG sensors can be used for the monitoring of strain and/or temperature at known punctual/specific locations in an optical fiber network. Strain wise, a physical elongation of the sensor induces the grating period and the photoelastic effect to change. As for a temperature sensor, the thermal expansion induced in the fiber material, associated to the thermal dependence of the fiber refractive index, induces a wavelength shift in the perturbed grating [51].

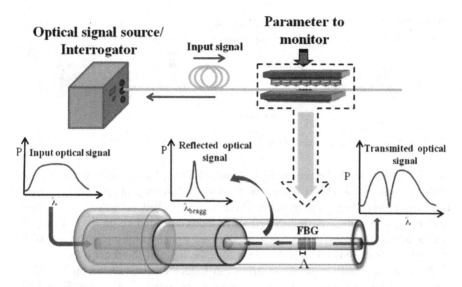

**Fig. 1.13** A sensing system based on FBGs

## *1.3.2  Fluorescence*

The development of fibers doped with materials with luminescent properties has been researched for a long time. It is primarily based upon the original work of Snitzer in the early 60s, when the use of neodymium doped fiber lasers and optical amplification configurations was proposed [53]. From this first initiative, a new and wide range of luminescent doped optical fiber has been developed [53]. As dopant, elements such as rare earths have been used mostly into silica-based fibers [54]. Different configurations are used for these sensors, but one of the most common configuration is the fiber end tip sensor. In this configuration, the optical signal propagates in the optical fiber into a probe of fluorescent material. The fluorescent signal is then coupled into the same fiber and guided to optical signal analyzer [2].

OFSs, employing fluorescent techniques by coupling crystalline materials into optical fibers, have been used in monitoring parameters such as temperature and pressure [51, 53–55].

## 1.4  Polarization

Polarization modulation based sensors rely on the fact that a given parameter (under analysis) can induce changes in the polarization state of the optical signal propagating in the optical fiber [56].

Polarization modulation sensors require the use of fibers, in which it is possible to preserve the polarization state along the propagation axis. Theoretically, the state

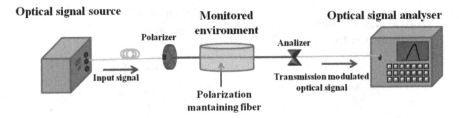

**Fig. 1.14** A setup of a polarization based OFS

of polarization should be preserved in standard fibers, but in practice, that is not completely achieved. This is mainly the reason why birefringent optical fibers, with two main polarization axes, are used. These axes are called fast axis (lower refractive index) and slow axis (higher refractive index), depending on the velocity at which the waves propagate through it. A polarized signal, travelling on the axis with the higher refractive index, travels slower than the signal in the faster axis [57].

As mentioned earlier, stress or strain can induce changes in the optical fiber refractive index (induced refractive index). Consequently, a phase difference between the different polarization directions axis is also induced in a phenomenon known as the photoelastic effect. Based on the induced refractive index effect, the external perturbations can be monitored by detecting the variations in the output polarization state [52]. Figure 1.14 shows the optical setup for the polarization based fiber optic sensor.

The basic configuration consists of monitoring the polarization of the input signal emitted by the optical source into a preferred axis of a polarization-preserving fiber (sensing section of the fiber). Under external perturbation (stress or strain), the phase difference between the two polarization states changes. As a result, the output polarization state is also altered according to the perturbation induced and so the external perturbation can be monitored by observing the output polarization state at the optical fiber end [58].

## 1.5   Distributed OFS

In a distributed optical fiber sensor (DOFS) system, the optical fiber is not only the transmission medium, but also a sensing component.

The scattering phenomenon is the base for the distributed optical fiber sensors (DOFS) configurations and can be described as the interaction between the optical signal and the propagation medium. Three different scattering processes are defined, namely: Brillouin, Rayleigh and Raman scattering [59]. The propagation of an electromagnetic wave in an optical fiber implies its interaction with fiber particles (atoms and molecules). This light/matter interaction creates a polarization dipole dependent of time, which can create a secondary electromagnetic wave, known as the light scattering [60]. In homogenous mediums, only a forward scattered signal is observed. Nevertheless, optical fibers are not a homogeneous medium, since its density and composition can vary. From the interaction between the optical signal and the fiber, some energy returns towards the light source originating the backscattering. The backscattering can then be used to evaluate not only the fiber properties, but also to retrieve information of the environmental effects altering the fiber. When strain and/or temperature changes are induced into the optical fiber, a modulation of the scattered optical signal within the fiber is created. DOFS is achieved by monitoring the variation of this modulated signal [59, 60].

DOFSs emerged with the development of a technique called Optical Time-Domain Reflectometer (OTDR) in the early 80s. In an OTDR, a short optical pulse is emitted into the optical fiber, while a photo detector analyzes the amount of the signal that is backscattered with the pulse propagation along the fiber [60]. The time difference of the backscattered signal of the input pulse is proportional to the length that the pulse has traveled. This principle is the base for OTDR phenomenon. For a launched pulsed optical signal with a pulse width of $\tau$, the location of the temperature or stress modulated optical signal change along the optical fiber can be assessed by the time delay of the speed of light, $c$, with the location accuracy being known as spatial resolution, $\Delta Z$, which can be obtained by [60]:

$$\Delta Z = \frac{\tau c}{2n_{eff}} \tag{1.5}$$

where $n_{eff}$ is the effective refractive index of the optical fiber. Typically, a distributed sensor can replace many point sensors, rendering it very cost effective solution. Also, it is logistically efficient (regarding weight and size), since it only requires one fiber and one optical monitor equipment for emitting and receiving the signal. This sensing mechanism can provide accurate local changes in temperature, stress, vibration, etc. along all the length extension of the fiber, making this DOFS a powerful solution for structural monitoring of civil or aerospace structures [60]. There are three main types of OTDR sensing mechanism, which are based on Brillouin [61], Rayleigh [62], and Raman [63] scattering, as detailed in the following three subsections.

## 1.5.1  Brillouin

Brillouin scattering is a product of the stimulated acoustic vibrations in the optical fiber [61]. These vibrations generate a counter-propagating wave, known as Brillouin scattering wave. In order to ensure the energy conservation law, the frequency between the Brillouin scattering signal and the original signal pulse is frequency shifted. This frequency shift is dependent on the temperature and the longitudinal strain oscillations. This dependence makes it possible to monitor temperature and strain variations along the fiber length, by using the Brillouin scattering effect [61]. Brillouin optical time domain reflectometer (BOTDR), based on Brillouin scattering, was developed with the aim of enhancing the traditional OTDR's range [59, 60].

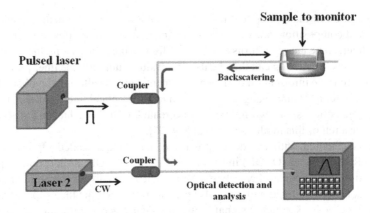

**Fig. 1.15** The sensing configuration based on the Brillouin scattering

The counter-propagating Brillouin scattering signal is normally a very weak signal; therefore, its detection requires some coherent techniques. In order to stimulate the Brillouin scattering (BS) process, a continuous wave (CW) is launched into the fiber from one end. The BS occurs, when the frequency difference between the pulse signal (from a pulsed laser) and the CW signal is equal to the Brillouin shift (resonance condition) [61]. In Fig. 1.15, a schematic diagram of a BS based sensing configuration is presented.

The first works on Brillouin-based DOFS were focused on temperature monitoring [64] and the optimization of its design on the Brillouin gain/loss mechanism enabled them to collect information over a length of 51 km (longest sensing length reported at that time) [59].

These long-distance distributed sensing mechanisms (typically up to 30 km BOTDR) can easily reach a sensitivity of 5 $\mu\varepsilon$; hence suitable for structural monitoring applications (large scale) [65]. Nonetheless, these systems are limited to a rough spatial resolution of 1 m; hence not adequate for a structural monitoring applications requiring higher resolution (less than 1 m) [59].

### 1.5.2  Rayleigh

Rayleigh scattering (RS) is characterized by the fact that the incident optical power is proportional to the scattered optical power, being therefore a linear scattering process with its origins in the density fluctuation of the optical fiber material [60]. Moreover, there is no energy exchange into the fiber core; therefore, the frequency of the scattered signal does not change, when compared to the frequency of the input signal [60, 66].

In Rayleigh-based DOFSs, the scattering itself is used mainly to locate and disclose propagation effects (attenuation, phase interference, polarization), which are the real sensing mechanisms that can be exploited in Rayleigh-based DOFS implementation [60].

When the scattering process occurs, part of the scattered signal is captured in the guided modes of the optical fiber and back propagates towards the fiber input. This "backscattered signal" is the one detected by Rayleigh-based DOFSs. Such signal is in general rather weak (several tens of dB less than the incident light) [66]. This fact represents the main technical difficulty in Rayleigh-based DOFSs.

Although the Rayleigh scattered signal may be weak (which is considered the main disadvantage of Rayleigh-based DOFSs), they are sensitive to several physical parameters besides temperature and strain, namely, relative humidity [67], concentration of chemicals [68, 69], radiation detection [70], vibration and intrusion [71], among others.

### 1.5.3 Raman

Raman distributed optical fiber temperature sensor system, which was reported in the mid-80s, is one of the most successful DOFS developed to date. This sensor employs the Raman scattering in the optical fiber to measure temperature values. The Raman backward scattering is modulated by the temperature gradient along the fiber axis, making it possible to determine the real-time temperature distribution in the fiber [63].

Raman scattering DOFS networks can be used in wide range of structures, including the monitoring of bridges, roads, dams, gas and oil pipelines, water supply systems. More importantly, their possible association with wired or wireless networks makes them suitable for the Internet of Things (IoT) scenario, for problem identification, assessment and control [72].

## 1.6 Summary

The use of optical fiber sensors in communications have been around for some time now. However, lately optical fibers have gained a lot of attention in the field of sensing, owing their success to a number of inherent characteristics of optical fiber sensors (OFSs). Inherently, optical fiber sensors are small in size, light weight, resilient to external harsh environment and immune to electromagnetic interference. There have been extensive research in the field of optical fiber sensing, resulting in multiple mechanisms of optical fiber sensing and many applications of OFSs in different fields. This chapter introduced the concept of optical fiber sensing, highlighting the advantages of using optical fibers in the field of sensing, while pointing out the limitations of the field. The chapter has gone through the multiple existing techniques (modulation/demodulation) used for sensing, listing the multiple types in each categories.

To sum up, optical fiber sensing can be divided into main categories, namely intrinsic and extrinsic, based on whether the perturbations occur within the fiber itself, or outside, in which situation the fiber is only used as the information carrier. Alternatively, OFSs can be divided into different categories based on the type of modulation/demodulation process and operating principal. Mainly, sensing using optical fiber can fall under one of the following categories: phase, frequency, intensity, or polarization sensor. OFSs operate through monitoring the value of those parameters, which is usually changed due to variations in the monitored external perturbations. Each of this category includes subcategories, which have been listed and detailed within the different subsections of the chapter. For each type, the advantages and limitations have been discussed.

This chapter can act as a starting guideline for readers new to the field of optical fiber sensing, to gain the required knowledge about the basics of sensing using optical fibers and to understand the different techniques used in OFSs. Through this chapter, the reader can obtain information with regards to which type of OFS to use or investigate further.

Having understood the basics and principles of optical fiber sensing and gotten familiar with the different existing categories of OFSs, the brief introduces different methods for the fabrication of each category in Chap. 2. The fabrication methods are detailed, while highlighting advantages and limitations of each fabrication method. Chapter 3 moves one step forward towards discussing different methods used to achieve lower-cost production methods of Silica OFSs. Chapter 4 complements Chap. 3, by introducing Plastic/Polymer Optical Fiber (POF), as an economical alternative to Silica Optical Fibers. Both Chaps. 3 and 4 represent a vital step towards producing low-cost OFSs, suitable for Internet of Things (IoT) applications. Chapter 5 then introduces the reader to the concept of IoT and its applications. The chapter provides a brief overview of the IoT concept and shows the importance of sensing within IoT. The chapter then concludes by presenting different fields, where OFSs have been used or would represent an added value.

# References

1. Ghatak, Ajoy, and K. Thyagarajan. *An introduction to fiber optics*. Cambridge university press, 1998.
2. Fidanboylu, K. A., and H. S. Efendioglu. "Fiber optic sensors and their applications." *5th International Advanced Technologies Symposium (IATS'09)*. Vol. 6. 2009.
3. Liaw, S-K., et al. "Pump efficiency improvement of a C-band tunable fiber laser using optical circulator and tunable fiber gratings." *Applied optics* 46.12 (2007): 2280-2285.
4. Wang, Yiping. "Review of long period fiber gratings written by CO2 laser."*Journal of Applied Physics* 108.8 (2010): 081101.
5. Zhu, Yinian, Zonghu He, and Henry Du. "Detection of external refractive index change with high sensitivity using long-period gratings in photonic crystal fiber." *Sensors and Actuators B: Chemical* 131.1 (2008): 265-269.
6. Miao, Yin-ping, and Bo Liu. "Refractive index sensor based on measuring the transmission power of tilted fiber Bragg grating." *Optical Fiber Technology* 15.3 (2009): 233-236.

7. Mohanty, Lipi, et al. "Fiber grating sensor for pressure mapping during total knee arthroplasty." *Sensors and Actuators A: Physical* 135.2 (2007): 323-328.

8. Mishra, Vandana, et al. "Fiber grating sensors in medicine: Current and emerging applications." *Sensors and Actuators A: Physical* 167.2 (2011): 279-290.

9. Grattan, K. T. V., and T. Sun. "Fiber optic sensor technology: introduction and overview." *Optical Fiber Sensor Technology*. Springer US, 2000. 1-44.

10. Krohn, D. A. "Fiber Optic Sensors, fundamentals and applications, 2000."*Instrument Society of America*.

11. Lee, Byeong Ha, et al. "Interferometric fiber optic sensors." *Sensors* 12.3 (2012): 2467-2486.

12. Maria de Fátima, F. Domingues, et al. "Liquid hydrostatic pressure optical sensor based on micro-cavity produced by the catastrophic fuse effect."*IEEE Sensors Journal* 15.10 (2015): 5654-5658.

13. Antunes, Paulo, et al. "Optical fiber sensors for static and dynamic health monitoring of civil engineering infrastructures: Abode wall case study."Measurement 45.7 (2012): 1695-1705.

14. Alberto, Nélia, et al. "Characterization of Graphene Oxide Coatings onto Optical Fibers for Sensing Applications." Materials Today: Proceedings 2.1 (2015): 171-177.

15. Leitão, Cátia, et al. "Central arterial pulse waveform acquisition with a portable pen-like optical fiber sensor." Blood pressure monitoring 20.1 (2015): 43-46.

16. Casas, Joan R., and Paulo JS Cruz. "Fiber optic sensors for bridge monitoring." *Journal of bridge engineering* 8.6 (2003): 362-373.

17. Anwar Zawawi, Mohd, Sinead O'Keffe, and Elfed Lewis. "Intensity-modulated fiber optic sensor for health monitoring applications: a comparative review."*Sensor Review* 33.1 (2013): 57-67.

18. Lau, Doreen, et al. "Intensity-modulated microbend fiber optic sensor for respiratory monitoring and gating during MRI." *IEEE Transactions on Biomedical Engineering* 60.9 (2013): 2655-2662.

19. Ilev, Ilko K., and Ronald W. Waynant. "All-fiber-optic evanescent liquid level and leak sensor." *Lasers and Electro-Optics, 1999. CLEO'99. Summaries of Papers Presented at the Conference on*. IEEE, 1999.

20. Warren-Smith, Stephen C., et al. "Temperature sensing up to 1300° C using suspended-core microstructured optical fibers." *Optics express* 24.4 (2016): 3714-3719.

21. Layeghi, Azam, Hamid Latifi, and Orlando Frazao. "Magnetic field sensor based on nonadiabatic tapered optical fiber with magnetic fluid." *IEEE Photonics Technology Letters* 26.19 (2014): 1904-1907.

22. Karabchevsky, Alina, P. Hua, and J. S. Wilkinson. "Simple evanescent field sensor for NIR spectroscopy." (2013).

23. Lee, Byeong-Ha, et al. "Specialty fiber coupler: fabrications and applications." *Journal of the Optical Society of Korea* 14.4 (2010): 326-332..

24. Rao, Yun-Jiang, et al. "Micro Fabry-Perot interferometers in silica fibers machined by femtosecond laser." *Optics express* 15.21 (2007): 14123-14128.

25. Kim, M. J., et al. "Simultaneous measurement of temperature and strain based on double cladding fiber interferometer assisted by fiber grating pair."*IEEE Photonics Technology Letters* 20.15 (2008): 1290-1292.

26. Ding, Jin-Fei, et al. "Fiber-taper seeded long-period grating pair as a highly sensitive refractive-index sensor." *IEEE Photonics Technology Letters* 17.6 (2005): 1247-1249.

27. Lim, Jong H., et al. "Mach–Zehnder interferometer formed in a photonic crystal fiber based on a pair of long-period fiber gratings." *Optics Letters* 29.4 (2004): 346-348.

28. Choi, Hae Young, Myoung Jin Kim, and Byeong Ha Lee. "All-fiber Mach-Zehnder type interferometers formed in photonic crystal fiber." *Optics Express* 15.9 (2007): 5711-5720.

29. Nguyen, Linh Viet, et al. "High temperature fiber sensor with high sensitivity based on core diameter mismatch." *Optics express* 16.15 (2008): 11369-11375.

30. Yuan, Li-bo, Li-min Zhou, and Jing-sheng Wu. "Fiber optic temperature sensor with duplex Michleson interferometric technique." *Sensors and Actuators A: Physical* 86.1 (2000): 2-7.

31. Kashyap, Raman, and B. Nayar. "An all single-mode fiber Michelson interferometer sensor." *journal of Lightwave Technology* 1.4 (1983): 619-624.
32. Tian, Zhaobing, Scott SH Yam, and Hans-Peter Loock. "Single-mode fiber refractive index sensor based on core-offset attenuators." *IEEE Photonics Technology Letters* 20.16 (2008): 1387-1389.
33. Park, Kwan Seob, et al. "Temperature robust refractive index sensor based on a photonic crystal fiber interferometer." *IEEE Sensors Journal* 10.6 (2010): 1147-1148.
34. Yuan, Libo, Jun Yang, and Zhihai Liu. "A compact fiber-optic flow velocity sensor based on a twin-core fiber Michelson interferometer." *IEEE sensors journal* 8.7 (2008): 1114-1117.
35. Islam, Md Rajibul, et al. "Chronology of Fabry-Perot interferometer fiber-optic sensors and their applications: a review." *Sensors* 14.4 (2014): 7451-7488.
36. Tsai, Woo-Hu, and Chun-Jung Lin. "A novel structure for the intrinsic Fabry-Perot fiber-optic temperature sensor." *Journal of Lightwave Technology* 19.5 (2001): 682.
37. Rao, Yun-Jiang. "Recent progress in fiber-optic extrinsic Fabry–Perot interferometric sensors." *Optical Fiber Technology* 12.3 (2006): 227-237.
38. Antunes, Paulo FC, et al. "Optical fiber microcavity strain sensors produced by the catastrophic fuse effect." *IEEE Photonics Technology Letters* 26.1 (2014): 78-81.
39. Wei, Tao, et al. "Miniaturized fiber inline Fabry-Perot interferometer fabricated with a femtosecond laser." *Optics letters* 33.6 (2008): 536-538..
40. Tafulo, Paula AR, et al. "Intrinsic Fabry–Pérot cavity sensor based on etched multimode graded index fiber for strain and temperature measurement." *IEEE Sensors Journal* 12.1 (2012): 8-12.
41. Wang, Zhuang, et al. "Multiplexed fiber Fabry–Perot interferometer sensors based on ultrashort Bragg gratings." *IEEE Photonics Technology Letters* 19.8 (2007): 622-624.
42. Majchrowicz, Daria, and Marzena Hirsch. "Fiber optic low-coherence Fabry-Perot interferometer with ZnO layers in transmission and reflective mode: comparative study." *Saratov Fall Meeting 2015*. International Society for Optics and Photonics, 2016.
43. Zhao, Jia-Rong, et al. "High-resolution and temperature-insensitive fiber optic refractive index sensor based on fresnel reflection modulated by Fabry–Perot interference." *Journal of lightwave technology* 28.19 (2010): 2799-2803.
44. Domingues, M. Fátima, et al. "Cost effective refractive index sensor based on optical fiber micro cavities produced by the catastrophic fuse effect."*Measurement* 77 (2016): 265-268.
45. Choi, Hae Young, et al. "Miniature fiber-optic high temperature sensor based on a hybrid structured Fabry–Perot interferometer." *Optics letters* 33.21 (2008): 2455-2457.
46. Domingues, M. F., et al. "Enhanced sensitivity high temperature optical fiber FPI sensor created with the catastrophic fuse effect." Microwave and Optical Technology Letters 57.4 (2015): 972-974.
47. Gong, Huaping, et al. "Curvature sensor based on hollow-core photonic crystal fiber sagnac interferometer." *IEEE Sensors Journal* 14.3 (2014): 777-780.
48. Ma, Jun, Yongqin Yu, and Wei Jin. "Sagnac interferometer based stable phase demodulation system for diaphragm based acoustic sensor.". *Fifth Asia Pacific Optical Sensors Conference*. International Society for Optics and Photonics, 2015.
49. Han, Tingting, et al. "Unique characteristics of a selective-filling photonic crystal fiber Sagnac interferometer and its application as high sensitivity sensor." *Optics express* 21.1 (2013): 122-128.
50. Shao, Li-Yang, et al. "Sensitivity-enhanced temperature sensor with cascaded fiber optic Sagnac interferometers based on Vernier-effect." *Optics Communications* 336 (2015): 73-76.
51. Grattan, K. T. V., and T. Sun. "Fiber optic sensor technology: an overview."*Sensors and Actuators A: Physical* 82.1 (2000): 40-61.
52. Lee, Byoungho. "Review of the present status of optical fiber sensors."*Optical fiber technology* 9.2 (2003): 57-79.
53. Wang, Xu-Dong, and Otto S. Wolfbeis. "Fiber-optic chemical sensors and biosensors (2008–2012)." *Analytical chemistry* 85.2 (2012): 487-508.

54. Wang, Xu-dong, Otto S. Wolfbeis, and Robert J. Meier. "Luminescent probes and sensors for temperature." *Chemical Society Reviews* 42.19 (2013): 7834-7869.
55. Xu, Wei, et al. "Optical temperature sensing through the upconversion luminescence from Ho 3+/Yb 3+ codoped CaWO 4." *Sensors and Actuators B: Chemical* 188 (2013): 1096-1100.
56. Woliński, T. R. "Polarization in optical fibers." *Acta Physica Polonica A* 95.5 (1999): 749-760
57. Song, Ningfang, et al. "Structure optimization of small-diameter polarization-maintaining photonic crystal fiber for mini coil of spaceborne miniature fiber-optic gyroscope." *Applied optics* 54.33 (2015): 9831-9838
58. Piliarik, Marek, et al. "Surface plasmon resonance sensor based on a single-mode polarization-maintaining optical fiber." *Sensors and Actuators B: Chemical* 90.1 (2003): 236-242.
59. Barrias, António, Joan R. Casas, and Sergi Villalba. "A Review of Distributed Optical Fiber Sensors for Civil Engineering Applications." *Sensors* 16.5 (2016): 748.
60. Bao, Xiaoyi, and Liang Chen. "Recent progress in distributed fiber optic sensors." *Sensors* 12.7 (2012): 8601-8639.
61. Galindez-Jamioy, Carlos Augusto, and José Miguel López-Higuera. "Brillouin distributed fiber sensors: an overview and applications." *Journal of Sensors* 2012 (2012).
62. Zhou, Da-Peng, et al. "Distributed temperature and strain discrimination with stimulated Brillouin scattering and Rayleigh backscatter in an optical fiber."*Sensors* 13.2 (2013): 1836-1845.
63. Bolognini, Gabriele, and Arthur Hartog. "Raman-based fibre sensors: Trends and applications." *Optical Fiber Technology* 19.6 (2013): 678-688.
64. Kurashima, Toshio, Tsuneo Horiguchi, and Mitsuhiro Tateda. "Distributed temperature sensing using stimulated Brillouin scattering in optical silica fibers." *Optics Letters* 15.18 (1990): 1038-1040.
65. Uchida, Shun, Eyal Levenberg, and Assaf Klar. "On-specimen strain measurement with fiber optic distributed sensing." *Measurement* 60 (2015): 104-113.
66. Palmieri, Luca, and Luca Schenato. "Distributed optical fiber sensing based on Rayleigh scattering." *The Open Optics Journal* 7.1 (2013).
67. Lenke, Philipp, et al. "Distributed humidity sensing based on Rayleigh scattering in polymer optical fibers." *(EWOFS'10) Fourth European Workshop on Optical Fibre Sensors*. International Society for Optics and Photonics, 2010.
68. Chen, Kevin P. "Ultrafast Laser Enhanced Rayleigh Scattering Characteristics in D-Shaped Fibers for High-Temperature Distributed Chemical Sensing." *Bragg Gratings, Photosensitivity, and Poling in Glass Waveguides*. Optical Society of America, 2016.
69. Gao, Zhong Feng, et al. "Detection of mercury ions (II) based on non-cross-linking aggregation of double-stranded DNA modified gold nanoparticles by resonance Rayleigh scattering method." *Biosensors and Bioelectronics* 65 (2015): 360-365.
70. Rizzolo, S., et al. "Radiation Characterization of Optical Frequency Domain Reflectometry Fiber-Based Distributed Sensors." *IEEE Transactions on Nuclear Science* 63.3 (2016): 1688-1693.
71. Peng, Fei, et al. "Ultra-long high-sensitivity Φ-OTDR for high spatial resolution intrusion detection of pipelines." *Optics express* 22.11 (2014): 13804-13810.
72. Zhang, Zaixuan, et al. "Recent progress in distributed optical fiber Raman photon sensors at China Jiliang University." *Photonic Sensors* 2.2 (2012): 127-147

# Chapter 2
# Silica Optical Fiber Sensors Production Methods

The concept of OFS is not new and has been around since the 1960s, but OFS has made the big leap to practical implementation, only thanks to the advances in modern low-loss optical fibers, although some trails using low-loss optical fibers were shown in 1970s [1, 2]. The field of OFS owes its advancement to the progress achieved in the general field of optical fiber and optoelectronic instrumentation. Specifically, the advancement in the field of optical fiber communication systems—such as viable key components: light sources and photo detectors—has helped bringing the costs of OFS systems significantly down. The low cost of sensors is one of the main drivers for the interest in the concept of optical fiber sensing and will be a key issue in the argument of adopting OFS in the concept of IoT, with a wide range of applications [3–5].

Having introduced the concept of optical fiber sensing and the different types of available optical sensing techniques (i.e. modulation/demodulation) in the previous chapter, this chapter moves to the details of the different production methods of Silica OFSs.

The chapter discusses the production methods of the five different types of sensors (previously introduced in Chap. 1), namely Intensity sensors, Phase Interferometer sensors, Wavelength sensors, Polarization sensors, and Distributed Optical Fiber sensors. Each type of sensors will be described in a separate section, with subsection describing the production methods and another subsection listing the most typical application of such type of sensors. The sections will further address the complexity of the production process, analyzing and comparing the sensitivity and accuracy of different types of manufactured sensors.

This chapter can be used as guidelines for readers, who are trying to decide which production method to use for their applications and requirements.

© The Author(s) 2017
M.F.F. Domingues, A. Radwan, *Optical Fiber Sensors for IoT and Smart Devices*,
SpringerBriefs in Electrical and Computer Engineering,
DOI 10.1007/978-3-319-47349-9_2

## 2.1   Intensity Sensors

The first type of sensors discussed is the Intensity-modulated sensors, which is amongst the earliest OFSs studied. Basically, the concept of Intensity-modulated sensors depends on modulating the intensity of light passing through the fiber by various means including bending, reflectance or changing the medium. The attractive attributes of such type of sensors are its ease of fabrication, simple detection method and low-cost [4, 6, 7].

### *2.1.1   Production Methods*

In the previous chapter, different techniques of intensity-modulated based fiber sensors have been presented, such as fiber displacement, transmission loss, reflective, and evanescent field. This subsection will discuss the production process of those different techniques.

The first technique is fiber displacement sensor, in which two optical fibers are connected in series. The light source and the detector are connected at either side of the optical fiber. When any movement or displacement, either in transverse, longitudinal angular or differential direction occurs between the two fibers, the light intensity received by the detector changes [8], as shown in Fig. 2.1.

In a reflective intensity sensor, the fibers for the input and output optical signal are placed in parallel configuration with each other (in a two fiber parallel coupling configuration), while the target object is placed in a close distance to the fiber ends, as illustrated in Fig. 1.7a of Chap.1. The target object reflects the light depending on the distance from the fiber end and the interaction of the reflective material with the incident optical signal. The optical fiber connected to the optical signal analyzer (OSA) detects the changes of the reflected optical signal [4, 8].

Another intensity sensor can be built based on the modulation of the transmitted light by the optical fiber, by inducing a bend on it. The measured parameter causes a bending in the fiber, which results in changes in the light intensity of the output signal. Any physical action able to induce or change a bend in the optical fiber can be monitored using this technique. There are two types of fiber bending, namely Microbend and Macrobend.

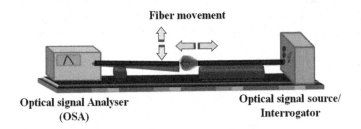

**Fig. 2.1** Schematic diagram of a displacement intensity sensor

**Fig. 2.2** U-bend sensing
configuration based on
macrobending

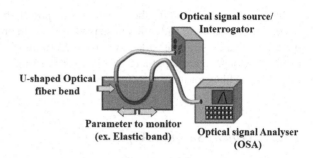

Microbend based sensors are used in applications where the measured parameter such as strain, pressure, or force can be mechanically attached and related to the movement of platforms that can induce microbends in the optical fiber [9]. The sensing fiber is placed in between teeth-like deformers to trigger the optical fiber loss. In that way, the sensor probe senses the measured parameter, which is induced by decreasing the distance between the upper and lower teeth, resulting in a series of bend formation along the length of the fiber [9]. The continuous movement of the grooved platform can be linearly related to the optical signal loss in the output fiber end. This mechanism was illustrated in Fig. 1.5 in Chap. 1.

The second type of bending is the macrobending technique. Contrary to microbending, no deformer is required to change light intensity in the macrobending process. In macrobending, the fiber is shaped into U-shaped, round loop(s), sinusoidal shape or other comparable bending formations, which permit attenuation in light intensity, as represented in Fig. 2.2.

The fiber sensing section is then placed in contact with measured parameter, which by altering the radius of the sensing fiber will induce changes in the light intensity that is received by the light detector. The change in the intensity of the output optical signal will be a function of the value shift of the parameter [8, 10].

The last type of intensity sensor is the evanescent field sensors, which is a phenomenon characterized by the fact that the optical signal along the fiber is not totally trapped in its core (for incidence angles near to the critical angle), but some signal leakage can occur into the surrounding medium, more specifically into the cladding [11].

The design of optical fiber is made for the optimum signal propagation with minimum loss possible. Nevertheless, when sensing is involved, it is essential that the optical signal interacts with the surrounding environment, in order to monitor its parameters.

To boost the interaction of the optical signal propagating in the fiber with its boundary, the evanescent field of the transmitted light should be exposed, which can be achieved by reducing or removing part of the optical fiber cladding. Once the cladding is removed, and the substrate is placed in contact with the optical fiber, the parameters under analysis can be detected through the leakages created by the absorption of the evanescent field, which modulates the optical signal intensity [3, 11, 12].

In this simplistic design, the penetration depth of the evanescent field is not enough for an accurate and efficient sensing [3, 12]. In order to optimize this sensing mechanism, research efforts have proposed solutions such as the use of macrobending, optical signal input angle shift, fiber tapering or the use of special fibers like D-shape fibers or microstructure fibers [13–17].

### 2.1.2   Typical Sensing Applications

Optical fibers are intrinsically safe for human use in the field of healthcare, due to their chemical inertness and non-toxicity, in addition to their dielectric nature and tiny size, which make them suitable for miniaturized devices or embedment into textile materials for sensing. Such characteristics of optical fibers have motivated extensive research of using high sensitive intensity based OFSs in physical, structural, biological, chemical and biomedical fields [4].

For instance, microbend OFSs have been explored, with good performance, for the monitoring of parameters such as pressure, temperature, strain, vibration, acceleration, displacement, humidity, pH, and in the monitoring of breath rate and respiratory/body movement [7, 18–21]. The main advantage of microbend OFSs over other OFS types is their design simplicity and low production costs.

In the rest of this subsection, different specific examples of health monitoring applications, which have been discussed in literature, will be presented, highlighting sensing configuration and advantages.

Low back pain (LBP) is one of the main health issues associated with the lumbar spine (part of the spinal column between the thorax and sacrum), and its diagnosis and rehabilitation procedure requires accurate assessment of the lumbar spine condition. For such precise diagnosis, Dunne et al. suggested using OFS for monitoring the lumbar spine in [8, 21]. They used an optical fiber bending sensor, which was attached to a wearable textile outer surface allowing the sensor to move vertically. They placed the optical signal source and a detector at either end of the fiber. The intensity of the output optical signal was modulated according to the degree of the bend in the fiber, which is directly correlated to the curvature of the lumbar spine [21].

Other approaches, based on intensity modulated, use a series of coupled fibers along the spinal, which modulates the spinal angle movements by monitoring the light intensity output [22]. Also using the coupling efficiency between two optical fibers, it is possible to monitor the upper limbs. The misalignment between two coupled fibers can be related to the angle of the joint movement [8].

Similarly, for the monitoring of the lower limb, different sensing solutions have been used. For instance, Bilro et al. presented an optical fiber technology based on the microbending effect to monitor the gait movements. In this approach, the fiber is placed along the knee joint in vertical alignment, while the optical signal source and the output detector are placed at different ends of the fiber. The knee movement is then measured by recording the light attenuation, which occurs due to direct fiber bending [23]. This technology is neither as bulky as similar inertial gait sensors, nor as expensive as video monitoring approaches [8].

OFSs are also applied for another health monitoring application, namely respiratory monitoring, where macrobending intensity modulation effect is used [7, 24–26].

In those approaches, configurations such as U-shape bent fiber or fiber shaped in sinusoidal form with a concentrated distribution, imbedded in a textile elastic belt (to be placed at the patient thoracic area), are used to monitor the respiration rate associated to the abdomen movement [24, 25].

Nevertheless, this approach still presents some drawbacks, such as uneven force distribution, un-proportionality between the bending loss and the number of loops in fiber and the fiber bending radius dependence on the optical signal input wavelength [26].

From a signal point of view, heart rate is similar to respiration rate, except for the frequency. The frequency of human heart beat lies in the range of 0.7–1.8 Hz (40–110 beats/min), while the respiration beat is much lower lying in the range of 0.2–0.5 Hz (12–30 beats/min) depending on the person age. Having such similarity, same (or at least very similar) sensing methods can be used for heart rate monitoring [8].

Finally, it is worth mentioning that the type of fiber has significant effects on the cost, performance and other handling concerns for the sensor implementation. One issue, related to the macrobending technique described in the applications mentioned above, is the difficulty to distinguish between normal and abnormal bending of fiber during the monitoring process. One possible solution is to implement the optical time-domain reflectometry (OTDR), which is able to monitor the loss distribution along the fiber length [8].

Besides its biomedical applications, intensity based optical fiber sensors also paved their way into other differentiated fields, such as the Structural Health Monitoring (SHM).

In the SHM field, this technology is applied for the monitoring of buildings, bridges, tunnels, wind turbines, railway, geotechnical and highway structures, pipelines, among others [27, 28].

Also, chemical parameters and solution elements can be monitored based on intensity OFSs. For instance, a flexible and accurate evanescent wave technology based on the cladding lateral polish of the optical fiber with low-cost light sources has been presented for pH, relative humidity and temperature monitoring [11].

## 2.2 Phase Interferometer Sensors

The second type of OFSs, to be discussed in this chapter, is the Phase-modulated sensors, which are also known as interferometric sensors. Phase Interferometer sensors are built based on the phase difference of coherent light, which travels through two different paths, either in one same fiber or different fibers. Interferometer

Optical Fiber Sensors (IOFSs) are characterized by high-sensitivity, since they are able to respond to the smallest changes in the external measured feature or parameter. Phase interferometric fiber sensors cover multiple types including Mach–Zehnder, Michelson, Fabry–Perot, and Sagnac interferometers [29].

The length of the optical path of each arm of the IOFS, which is a derivate of the fiber refractive index and the fiber geometrical length, can be influenced by environmental physical parameters. This optical path length variation can be precisely measured by optical fiber interferometry. By measuring the phase difference, the change in the environmental physical parameter can be quantified. This is the principal concept of IOFS operation [30].

## 2.2.1  Production Methods

It is worth mentioning that the nature of the targeted application affects the production of OFSs in general, and IOFSs are no different; hence IOFSs are produced using different techniques, either for discrete, quasi-distributed, or distributed fiber sensing applications. In applications as intruder detectors, where IOFS are produced using distributed or quasi-distributed monitoring, commercial optical fiber can be used. This commercial optical fiber is basically the same as the one used for optical fiber communications. In many applications such as for temperature, fiber optical gyroscope (FOG) or strain measurements, no special optical fiber is required [30].

Not only the optical fiber depends on the application, but the jacketing and cladding of the OFS have to be designed to suit the specific application. For instance, for strain sensing, the optical fiber cable structure must be resistant to temperature shift and block the heat transfer to the optical fiber sensing element, in order to prevent the temperature influence in the measured strain values [29].

On the other hand, in other applications (bio chemical sensing) based on refractive index change or other different spectroscopy methods, the cladding of the fiber sensor should be completely removed, so the substrate maintains a direct contact with the sensing element. Different methods for removing fiber cladding have been used, including chemical etching by hydrogen fluoride acid and fiber tapering by $CO_2$ laser or hydrogen torch [31].

As previously introduced in Chap. 1, different types of IOFS include Fabry–Perot, Mach–Zehnder, Michelson, and Sagnac principle based sensors. Owing to their several advantages such as easy alignment, high coupling efficiency, and high stability, IOFSs have been widely investigated for micro-scale and in-line applications [30]. Within the in-line configuration, the sensing element can be intrinsic or extrinsic to optical fiber.

For Extrinsic Fabry-Perot Interferometer (EFPI) fabrication, the cavity is formed outside the optical fiber, by external reflective surfaces. The advantage of such configuration is its easy production method with low-cost equipment; nevertheless, they may present some drawbacks, such as packaging, accurate alignment and low coupling efficiency [29]. To overcome such low coupling efficiency issue, Photonic

Crystal Fibers (PCFs) are used. An FPI cavity can be formed by inducing the air holes for the fiber to collapse, by electric discharge or laser incidence. The collapse of the air holes can easily create a lens at the PCF end, which acts as a reflective surface [30, 32, 33].

In contrast, in the case of Intrinsic Fabry-Perot Interferometers (IFPIs), the mirrors are placed inside the optical fiber itself. IFPIs can become expensive, since they require high-cost equipment for fabrication [29]. Nevertheless, new low-cost solution has also been reported for the production of inline FPI microcavities, only by splicing a standard optical fiber with recycled optical fibers previously destroyed by the catastrophic fuse effect, which presents a considerable economical solution of optical fiber sensing [34, 35].

Among the interferometric OFSs, sensors can be based on a resonance or on a non-resonance mechanism. One sensor based on resonance is Sagnac resonators, which operate on the principle of constructive and destructive interference. The resonance frequency varies according to the changes in the monitored environmental parameters. The measurement of such resonance frequency is used to monitor the physical parameter under study. The output optical signal contains the amplitude or phase shift at a given frequency, which can provide the information of the monitored surrounding environment [29].

Alternatively, non-resonance IOFSs function is based on comparing the optical path difference (OPD) of the different arms of the optical fiber interferometer (Mach–Zehnder, Michelson based sensors). In such technique, at least two arms are used, while only one of those arms is exposed to the environmental variations. The other arm is placed isolated from the variation, working as the reference arm. The OPD can be easily measured by monitoring the changes in the output optical interference signal. Resonators phase is strongly influenced by the optical length change in the resonance frequency proximity. In order to enhance the sensitivity of a non-resonance IOF, the implementation of resonators in the sensing arm is highly advised [30].

### 2.2.2  Typical Sensing Applications

Different sensing configurations (from point, quasi-distributed or distributed sensing) are chosen based on the location of the required measurement. Point sensors are used when monitoring a physical parameter at a certain position, while distributed fiber sensors are the choice, when continuously measuring a sensed parameter along the length of the fiber. Finally, quasi-distributed sensors can monitor a physical parameter at finite positions, along the length of the optical fiber. Interferometric OFSs can be implemented in the three different configurations [36].

For a quasi-distributed sensing implementation, some multiplexing techniques are required to be implemented. Different multiplexing techniques can be used, namely frequency division multiplexing (FDM), time division multiplexing (TDM), code division multiplexing (CDM), wavelength division multiplexing (WDM), space division multiplexing (SDM), and hybrid multiplexing schemes [36–39].

The multiplexing schemes (to transmit the output data) are chosen considering features like the number of sensors to multiplex, the sensor frequency response, the dynamic range need, the financial cost, and economic impact [30].

All types of IOFSs are suitable for temperature monitoring. In Mach–Zehnder and Michelson IOF, the reference arm is kept at a given constant temperature, while a sensing arm is deployed at the position where temperature monitoring is required. In-line Mach–Zehnder and Michelson interferometers can also be used as integrated temperature sensors.

When using Sagnac interferometer for temperature sensing, the fiber of Sagnac ring is doped in order to achieve a high thermal expansion coefficient, which induces a high-birefringence shift [29, 30].

Another application, where interferometer sensors are used, is strain sensing. Strain usually induces a considerable variation of the optical fiber dimensions, which lead to strong shift of the refractive index of the fiber (both in the core and cladding) [29]. However, strain sensing OFSs sometime suffer the drawback of mutual sensitivity to temperature. But with the choice of an appropriate fiber protection or its composition, the effect of the variation in both parameters can be efficiently separated. Practically, it is possible to distinguish the different effects and hence produce sensors insensitive to one of the parameters (temperature or strain) [30]. The strain sensing OFSs are adequate to use for many applications including seismographs, hydrophones, geophones, structural health monitoring, etc.

An acoustic OFS, known as fiber hydrophone, is characterized by its high sensitivity and wide dynamic range when compared to conventional piezoelectric (PZT) ceramic sensors. Additionally, optical fiber hydrophones are immune to electromagnetic interference, do not require electronic devices at the monitoring set and can be deployed in a large number. Those features are essential for applications such as seismic exploration for oil reserves. Optical fiber hydrophones are, therefore, the answer to this quest, since they are able to multiplex a big number of sensor outputs on a single optical fiber [30, 40].

IOFSs are adequate to produce both types of strain sensors: in transmission and reflection configuration. Mach–Zehnder interferometers have been widely used in the early optical fiber hydrophone systems, specifically in transmission-type sensor [41]. In contrast, the Michelson configuration is more appropriate for building a reflective fiber hydrophone array. In such configuration, the induced light passes through the sensing arm twice; thus, providing twice the sensitivity of a Mach–Zehnder interferometer [41].

Alternatively, the Fabry–Pérot optical fiber hydrophone is considered the simplest transmission-type strain sensor, which possesses the advantages of being compact, but suffers from small dynamic range [41–44].

Due to OFSs ability to measure strain and temperature simultaneously through multiplexed interferometric, they offer various advantages in structural engineering, both for the quality control during construction process and also, thereafter for monitoring the health of buildings through time. IOFSs have been demonstrated for use in many important structures such as aircraft, marine vehicles, dams, and bridges. Moreover, IOFSs have been also implemented in the electric power industry for remote load monitoring of power transmission lines [27, 34, 35, 45].

## 2.3 Wavelength Sensors

Moving on towards one of the most commonly investigated type of OFSs, this section discusses wavelength-modulated OFSs. The concept of wavelength-modulated OFSs is built based on the idea of the variation in the propagating optical wavelength (spectral modulation) by an external monitored perturbation. Some of the common wavelength-based OFSs are Bragg grating sensors and fluorescence sensors.

Out of those wavelength modulated OFSs types, Bragg grating sensors are the most commonly used and are considered the most promising development technique [46].

Fiber Bragg gratings (FBGs) are written in optical fibers, by inducing periodic modifications in the optical fiber refractive index [46]. Such modulated periodic structure enables the propagating optical signal to be coupled into a backward propagating core mode, creating a reflection optical signal [46]. This reflected optical signal (reflected Bragg wavelength) is sensitive to a range of physical parameters; hence, FBG-based sensors can be used in a variety of applications [47, 48]. On the other hand, fluorescent-based OFSs are used for different physical and chemical monitoring, including temperature, humidity and viscosity [49].

In the following section, the different production methods of wavelength modulated OFSs are discussed, concentrating on FBGs development due to its popularity. These sensing elements attracted a widespread research level, which boosted its rapid development. Hence, multiple techniques have been elaborated, along with several demodulation techniques [47].

### 2.3.1 Production Methods

The discovery of FBG modulation had significant impact on the development of optical fiber sensing. Since they were first investigated for sensing purposes in 1989, FBG sensors have gained wide attention resulting in a variety of demodulation techniques for different monitored parameters and diverse applications [48]. Due to its popularity and widespread use in telecommunications and sensing application, the production methods of FBGs are discussed in more details compared to other OFSs techniques.

FBGs in general are very sensitive to any slight perturbation of the fiber in the grating region; hence they have been extensively investigated in the field of optical sensing. The information with regards to the variation in the measured parameter is wavelength encoded [50].

FBGs are usually functioning based on periodic perturbation of the refractive index along the optical fiber core longitudinal axis [47]. The grating structure is inscribed in the optical fiber to form a selective reflective structure for the wavelength, which satisfies the Bragg condition.

When an optical source is used to illuminate the FBG, only a narrow band spectral component, corresponding to the Bragg wavelength of the grating ($\lambda_{Bragg}$), is reflected, whereas all the rest of wavelengths outside the reflection band are allowed to pass [47, 50].

The fabrication of FBGs has passed through multiple phases to reach its current state. In 1981, Lam and Garside mentioned that the mechanism of the formation of FBGs was basically due to the interaction of ultraviolet (UV) light with defect states of the doped silica core [47, 51]. Later on, it was shown that any change in the refractive index could be induced in all germanium-doped optical fibers. This finding has opened new doors in the manufacture of FBGs, resurrecting the interest in its research activity [47].

However, it took almost a decade till another breakthrough in FBG inscription occurred. The technique popularly known as the external inscription technique of writing FBG, was demonstrated by Meltz et al. [47, 52].

In this approach, a 244 nm beam is used, which is split into two and recombined to create an interference pattern at the photosensitive fiber core. The interference pattern induces a punctual and permanent change in the core refractive index [52].

The period of the interference pattern and the refractive index change are determined by the angle between the two split beams, which can be adjusted to produce gratings according to the desired reflected Bragg wavelength [48, 52].

Alternatively, FBGs can be written using phase masks [53]. A Phase mask is made of period patterns usually etched onto fused silica [47]. In such configuration, when the UV radiation hits the phase mask, the zero-order diffracted beam is almost totally suppressed (less than 3% of the transmitted power), while the diffracted orders +1 and −1 are maximized (>33% of the transmitted power) [53]. Figure 2.3 shows the schematic setup used to write FBGs using a phase mask.

Another more flexible method for inscribing Bragg gratings into an optical fiber is the *point-by-point* method. This method uses an UV source, which is passed through a slit and focused on the fiber to induce a refractive index change at a certain location. The fiber is then shifted a distance "$\Lambda$", which matches the period of the Bragg condition [47, 54]. However, Point-by-point technique cannot be applied for first-order gratings, due to its submicron translation and tight focusing requirements. Basically, the length of the grating needs to be considerably small and research efforts have only managed to produce second and third-order gratings, like the work of Malo et al. [47, 54].

There are multiple methods for producing grating fringes in optical fiber, using different photosensitive mechanisms. The underlying photosensitivity mechanism is referred to as the *type of gratings*. The method used to create the grating fringes greatly affects the physical attributes of the produced grating, specifically temperature sensitivity and resistant to elevated temperatures. According to current literature, five types of FBG exist with different photosensitivity mechanisms, namely Type I, Type IA, Type IIA, Type II, and regenerative grating [47].

*Type I* regime depends on the monotonic increase in refractive index modulation of a grating using UV exposure. The growth dynamics of this type of grating can be described by a temporal evolution of the refraction index [55]. Although the inscrip-

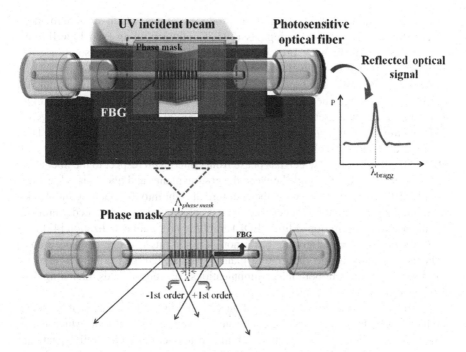

**Fig. 2.3** Set up typically implemented to inscribe FBG sensors in photosensitive optical fiber

tion process is quite simple in photosensitive optical fibers, Type I gratings present a low thermal stability [47]. For temperatures below 300 °C, the grating modulation remains constant, but it starts decaying for higher values of temperature.

*Type IA* gratings are basically a derivate of Type I gratings. They are simply produced through exposing a standard grating in hydrogenated germane-silicate fiber to a prolonged UV beam [56]. Type IA gratings have the lowest temperature coefficient, when comparing with other grating types; hence, they are ideally adequate for temperature compensating in sensor applications.

*Type IIA* FBGs are similar to Type I gratings, when it comes to spectral characteristics. The gratings require a long process for inscription, which is usually done following Type I grating inscription [38]. Contrary to Type I, Type IIA gratings are more stable at temperatures up to 500 °C [47].

*Type II* Bragg gratings are produced using excimer lasers, with very high pulse (>0.5 J/cm²) [57]. The concept is based on the idea that a sudden sharp increase in the refraction index happens, once the power of UV light passes a certain threshold. Such sharp index change, known as Type II, results in the highest stability at extreme temperatures up to 800 °C [47].

The last type of gratings explained here are the Regenerated gratings. The Regenerated gratings can be produced by annealing gratings of Type I; hence the name Regenerated. The regeneration process should occur at temperatures higher than 500 °C. Regenerated gratings are more similar to Type II gratings and they can be repeatedly cycled to very high temperatures without degradation [47, 58]. Recent findings have shown that this method can produce gratings, which are able to operate at extreme temperatures, as high as 1295 °C, outperforming even Type II gratings. This performance presents an extremely attractive solution for ultrahigh-temperature applications [58].

Additionally, FBGs can be classified depending on both the grating periodicity and refractive index. The most common category of FBG is the uniform grating, which has a constant grating pitch. The grating period is typically in the range of 0.25–0.5 μm, with the light coupled into the backward propagating direction. This type of grating is an excellent temperature and strain sensing [47]. Another type of FBG is the tilted grating, which has phase fronts tilted with respect to the fiber axis. In such configuration, the angle between the grating planes and fiber axis is less than 90° [47, 59]. These gratings allow the coupling of light into the cladding into backward propagating radiation modes. Based on its characteristics, the tilted grating is usually used in erbium-doped fiber (EDF) amplifiers for gain equalization [47].

Moreover, the chirped grating has a non-periodic pitch, which displays a constant increase/decrease in the grating planes spacing. Chirped grating is used in multiple applications in both telecommunications and sensor fields, including dispersion compensation [47, 60].

Finally, to conclude the discussion of the production of FBG sensors, it has to be mentioned that FBG sensors have one basic and very important trait, which is the differentiation between the effects of strain and temperature. Generally speaking, temperature and strain have very similar effects on Bragg wavelength shift. It is very important to isolate the effects of temperature variation (when monitoring temperature) from effects of other physical parameters, specifically strain, and vice versa. This motivated many techniques to be developed over the years to discriminate the effect of strain and temperature. Two main categories have been defined in the industry for temperature-compensating methods, namely, extrinsic and intrinsic methods. As the name implies, extrinsic method use external material, which is usually combined with the grating to compensate the temperature, whereas obviously intrinsic methods use certain properties of the optical fiber itself [47].

## 2.3.2  Typical Sensing Applications

Having discussed the production processes of wavelength based OFSs and specifically FBGs, this sub-section goes over the most common sensing applications of wavelength-based OFSs. Wavelength-based OFSs, and specifically FBGs, are famous for strain, temperature and vibration sensing. Strain sensing is usually used to monitor other physical parameters.

The first application to be discussed and one of the most important uses of Bragg gratings is strain monitoring, whether static, quasi static, or dynamic.

For strain sensing, FBGs are written on the core of a conventional single-mode optical fiber. When applying strain along the longitudinal axis of the fiber at constant temperature, the Bragg wavelength shifts due to the change in the grating periodic spacing and the induced photoelastic variations in the effective refractive index of the optical fiber core [47].

FBG based sensors present a linear dependence between the strain and the reflected Bragg wavelength shift of the sensing element, within the elastic deformation limit of the fiber [47]. Also, the sensitivity of the produced FBG sensor is highly dependent on the photosensitivity of the fiber used in the production of the FBG, in addition to the fiber diameter [47].

Although strain is the direct measurable parameter, strain sensing FBGs can be employed to indirectly measure force, displacement, vibration, pressure, liquid level and flow dynamics, and magnetic field.

One major application of FBGs is vibration sensing, which is highly adequate for structural health monitoring (SHM) and damage detection [61–63]. These vibration sensing systems operate in two distinct themes: (a) Single-point system, which monitors localized region and (b) Quasi-distributed sensing system, which provides sensing capabilities along the whole length of an optical fiber [47]. Vibration sensing FBGs provide low-frequency and high-frequency response, which are adequate for different applications based on their nature. As for low-frequency response accelerometers, they are best suited for large structures such as bridges, towers, dams, etc., due to their low inherent frequencies. High-frequency response vibration sensors with high spatial resolution are widely used for crack detection of materials, where the frequency of such events could be elevated [61–64]. The concept of FBG vibration sensor is based on the fact that vibrations induce high-speed dynamic strain variations in the FBG sensing element; hence, by monitoring the shift of the Bragg wavelength the measuring of such vibrations is made possible [61–64].

As previously mentioned, FBGs are also highly sensitive to temperature and therefore they are widely deployed specifically in harsh environments due to their high resilience [47]. The thermal-induced changes in Bragg wavelength occur due to two factors. First, temperature changes result in the variation of the effective refractive index of the optical fiber core guided mode and second, the variations in temperature modify the grating period [47, 49].

Typical temperature responses are 6.8, 10, and 13 pm/°C for the Bragg wavelength of around 830, 1300, and 1550 nm, respectively [47].

Although mentioning FBGs always indicates permanent refractive index modulation in the fiber core, it was shown that extreme high temperature environments may cause the refractive index modulation to be erased. The maximum tolerated temperature recorded for conventional FBG sensor is around 600 °C, basically due to the weak bonds of germanium and oxygen of the fiber core material [47].

It is reported that thermal stability of the gratings is directly dependent on the composition of the fiber core used for the writing of the gratings. Despite the intolerance of conventional FBGs to extreme high temperatures, special compositions of optical fibers dopants have been proposed to create FBG temperature sensors sustainable up to 1000 °C [65–69]. Some examples of those successful efforts include gratings in fluorine-doped fibers, gratings written into N-doped fibers [65, 66], regenerated gratings [69], and the use of femtosecond IR lasers for the gratings inscription [47, 65–67].

Humidity is another critical physical parameter that is usually required to be monitored constantly in certain applications. This parameter, and its accurate control, is highly important in air conditioning maintenance for human comfort, fighting bacterial growth, and also in the in quality control process of certain products, etc. Humidity is defined as the amount of water in gaseous state present in the atmosphere of a certain environment. A more important term is the relative humidity (RH), which is defined as the ratio of the amount of water vapor actually existing in the environment to the maximum possible amount that the same environment can carry [68–70].

RH sensing using a polyimide-coated FBG was proposed by Kronenberg et al. [70]. The concept is based on the fact that the polyimide polymers, being hygroscopic, tend to absorb moisture and swell in damp atmosphere, as water molecules migrate into their material. The swelling of the polyimide coating is used as a stimulator to force strain on the FBG-containing fiber. Clearly, such strain modifies the Bragg conditions; thus varying the Bragg wavelength accordingly, which is monitored to keep track of the RH [47].

Optical fiber gas sensors have been designed to detect hazardous gases. Nowadays, Hydrogen is turning into an attractive alternative energy source for clean-burning engines and power plants, or even in critical applications, such as the Space Shuttle main engine. Unfortunately, the use of highly flammable liquid $H_2$ does not come without its own risks. Using such highly flammable fuel introduces safety hazards, due to its low explosive limit and rapid evaporation rate. Those risks attracted some efforts to design and propose OFSs to detect the leaks in hydrogen-fueled systems [47]. To control Hydrogen based processes, Butler and Ginley proposed a technique for sensing hydrogen leaks [71]. They built their design on the realization of the change in the elastic property of Palladium (Pd) during Hydrogen ($H_2$) absorption. Their model can be used to determine the axial strain in the fiber core. The sensing technique is based on tracking the induced mechanical stress in the palladium coating, as it absorbs $H_2$. The fiber is stretched by the stress induced in the palladium coating, thus expanding the grating period, which shifts the FBG Bragg wavelength [71].

## 2.4  Polarization Sensor

The concept of Polarization sensor is mainly based on the change of the state of polarization of the light wave propagating through the optical fiber. Such state change occurs due to the difference in the phase velocity of the two polarization components in a bi-refringent fiber. Different factors can vary the polarization properties of optical signal travelling in a fiber, including strain, pressure, stress and temperature. In polarimetric OFSs, the change in the state of polarization is monitored and used to access a specific sensing parameter [72, 73]. In the rest of this section, the details of the fabrication process of polarization OFSs are discussed, along with the most widely used applications.

## 2.4.1 Production Methods

As initially stated, polarimetric OFSs typically use high birefringent (HB) optical fibers, such as *bow tie* fiber, *panda* fiber, or elliptical core or polarization maintaining PCF fibers [74]. HBs are produced with high internal birefringence, which is achieved through engineering the core and/or cladding with an elliptical geometry, or by inducing an anisotropic stress into the cross-section of the fiber [74].

This physical structure of the fibers allows the phase velocities difference for the two orthogonally polarized modes to be big enough, to avoid the coupling between these two modes [74].

HB fibers, under external deformations, present the lowest order modal behavior, which is of special interest for sensing applications. However due to the high sensitivity of polarimetric OFSs to external parameters, such as strain and temperature, cross sensitivity is usually a major issue. As has been mentioned in the case of Phase Interferometer sensors, differentiation between sensitivity to different external factors has to be addressed during sensor design and application [74].

In a typical polarimetric sensor, polarized light is inserted at 45° into the main axes of a birefringent fiber, such that both polarization modes become equally excited. The polarizer analyzer (with a 90° relation to the input polarization state) converts the polarization state into intensity [72, 74]. Therefore, in such type of sensors, a change in polarization is translated into a variation in intensity. Using correlation between the change in the output intensity and the measured parameter, a polarimetric fiber sensor is produced for a variety of applications [72, 74].

## 2.4.2 Typical Sensing Applications

As for most OFSs, polarimetric OFSs can be implemented to measure strain, temperature, corrosion, pressure, among others [72, 73, 75, 76].

Additionally, polarimetric OFSs can further be used to measure current and voltage, using a variety of polarization-based effects, including optical activity, electro-optic effect and Faraday effect [72, 74].

Structural health monitoring (SHM) has always been a hot topic, due to the use of composite materials and the dangers associated with collapsed structures. Lately, with the progress in technology and the advances in Internet of Things (IoT) fields, specifically smart cities, the concept of the so-called "smart" composite structures with compact integrated sensors has become a possibility. Polarimetric OFSs perfectly fit the requirements for deployment in such structures [72].

In composite structures, the monitor of the thermal expansion behavior is of critical importance; hence monitoring such physical parameter is of critical necessity to avoid the debonding induced by differential thermal expansion stresses. A wide range of polarimetric sensors can be used by employing polarization maintaining (PM) and micro-structured highly bi-refringent (HB) polarization maintaining fibers. Such sensors are capable of sensing multiple parameters, such as temperature, strain, displacement and rotation [75, 77].

Temperature variation induces composite material thermal elongation, which is the result of a combination of multiple factors including thermal strain, caused by thermal expansion, induced by moisture absorption, or induced by residual strain [75, 77].

The composite material thermal elongation applies stress to the polymeric fiber sensors, resulting in a significant phase change. By measuring the phase shift and determining the strain sensitivity of the polymeric fiber sensors in constant temperature settings, the strain induced by the thermal elongation can be derived over the sample length [77].

Polarimetric OFSs are advantageous in SHM field, as one such sensor typically only employs a single high-birefringence (HB) fiber for sensing and returns the integrated strain over the sensing region [75]. When compared to 'point-based sensors', the HB fiber is sensitive to perturbations over its entire length. Data about damage at multiple points can be recorded; hence providing a more integral and better indication of the severity of damage compared to conventional point-based sensors [75]. Polarimeteric OFSs are also more flexible and easily distributed (can be implemented in surface or internally embedded in structures with minimum effect), lightweight, and with very high sensitivity, while effectively separating effects of external stimuli [72, 75].

## 2.5 Distributed Optical Fiber Sensors

Distributed sensing refers to the concept, where the fiber itself is the sensing element, and the sensing can be made along the whole length of the optical fiber. Therefore, the parameter under study can be continuously monitored along the overall length of the fiber as a function of time. Figure 2.4 demonstrates such application.

Brillouin, Rayleigh and Raman are three different types of scatterings, famous for the production of distributed optical fiber sensors (DOFSs). The three different scattering types are formed through the interaction of the optical signal carriers with local material properties shift like density, temperature and strain [78].

For instance, if an acoustic/mechanical wave passes through a fiber, a dynamic density change would be generated; such change may be affected by strain, temperature, vibration or birefringence. If changes in the light scattered along a fiber (frequency, phase, amplitude) can be detected, a distributed fiber sensor is then created to access localized oscillation of parameters such as strain, vibrations, temperature, or birefringence, with an action range varying from just few meters all the way up to one hundred kilometers. Such measurement can be realized in the time domain or frequency domain [78, 79]. The produced sensors can be used for disaster prevention in civil structures, through the continuous monitoring of bridges, pipelines, railroads, dams among others [80].

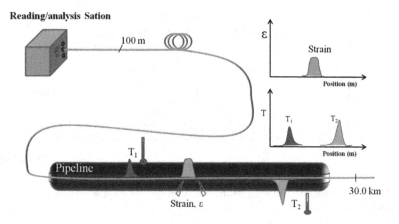

**Fig. 2.4** Distributed optical fiber sensing application

A distributed optical fiber sensor is required to return the information on the values of temperature, strain and vibration, from any location along an optical fiber, based on the light scattered along it [78, 80].

Hence its sensing capabilities are distributed over the whole length of the fiber. The challenge is to determine the value of the monitored parameter at exact points along the fiber length, with the required sensitivity and spatial resolution. DOFSs have met such requirement, by providing high performance distributed strain or temperature sensors, which are adequate for applications that cover large areas with relevant location accuracy [78].

For instance, Brillouin-based DOFS specifically provides strain resolution in a few micro-meters range over one meter (micro-strain) and temperature resolution reaching less than 1 °C [79]. Standard communication fibers can be easily embedded in civil structures (bridges, dams, power plants, buildings, airplanes, etc.), to monitor the internal wellbeing status of those structures [78].

## 2.5.1  Production Methods

As mentioned above, the distributed sensing is performed by connecting the scattering mechanisms with reflectometry techniques in either the time or frequency domain. When used in the time domain, it is referred to as optical time domain reflectometry (OTDR), whereas the frequency domain type is called optical frequency domain reflectometry (OFDR) [78].

Different scattering mechanisms can be invoked when coupling an electromagnetic wave into a silica wave-guide, as shown in Fig. 2.5.

Raman scattering can be produced using such method of coupling an electromagnetic wave into a silica wave-guide. Such method is capable of providing a precise measurement of temperature [81]. Such method of producing Raman scat-

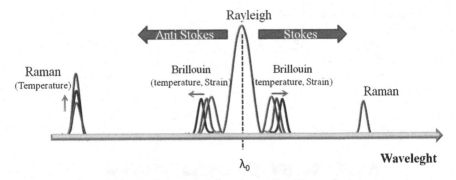

**Fig. 2.5** Different possible scattering mechanism observed over the optical signal propagation on an optical fiber

tering is called the spontaneous process and the sensing fiber can be either a single-mode or multimode optical fiber. By measuring the anti-Stokes scattered radiation, the Raman scattering presents itself as a perfect solution for distributed sensing of temperature [81]. This distributed sensing is also advisable for industrial applications requiring high precision, since it has very low sensitivity to other physical parameters such as pressure or strain, dodging its joint influence in the fiber.

Distributed sensing analysis is provided by a light change in the Optical Time Domain Reflectometry (OTDR) receiver to produce a band-pass filter able to separate the Stokes from the anti-Stokes components [78]. The temperature measurement is derived from the ratio between the Stokes and anti-Stokes components, meaning that simultaneous OTDR traces must be obtained for both components; hence a dual channel receiver should be used [79, 81]. Additionally, high transimpedance gain is required for such type of receivers and the bandwidth has to be constrained due to the low level of Raman backscattered signals [78].

The second type of scattering is the Brillouin scattering, which can be formed through the interaction of the incident optical signal and acoustic wave in the material [79]. The thermoelastic oscillation of the optical fiber core constituent molecules results in random density fluctuations, which propagate along the fiber in random directions as acoustic waves. The periodic modulation of density induces a refractive index modulation, which is typically small, but yet still causes diffraction of the incident optical signal [79].

The movement of the density modulation (along with the refractive index grating) at an acoustic velocity induces a frequency shift in the scattered light due to the Doppler Effect: the spontaneous Brillouin scattering. If the spontaneous scattering is intense enough, the incident and the scattered signals interfere, intensifying the acoustic waves. This constructive interference increases the strength of the refractive index grating, resulting in a high intensity of the scattered optical signal [80].

The acoustic wave propagation characteristics are directly related to the optical fiber material density, which in turn is affected by the influence of the medium and by parameters such as strain and temperature. Based on that principle, Brillouin scattering appears as a suitable solution to monitor parameters, such as strain and temperature. Sensing is performed by monitoring the Brillouin backscattered frequency, as the optical interference pulse propagates along the fiber. A major challenge in such production is to accurately access the Brillouin frequency shift. Brillouin scattering also suffers dual sensitivity to strain and temperature; hence there is the challenge of discriminating between the effects of both [78, 79, 80].

## 2.5.2 Typical Sensing Applications

DOFSs possess multiple characteristics that make them attractive for an array of various applications. First, reduced size sensor (fiber diameter is typically 150 μm) with a spatial resolution of centimeters and high precision in sensing parameters as temperature, vibration, strain, or birefringence is an attractive choice for smart precise applications in smart structures, material processing, and the characterization of optical materials and devices [80].

On another forefront, a wide range of sensing applications would profit from distributed sensing potential of spanning in the special scale up till six orders of magnitude [78, 80]. For instance, distributed temperature sensor (DTS) is one of the most sought-after applications in this domain [78, 81].

The world is currently seeing great progress in the area of renewable energy sources and the installation of smart grid systems, where power distribution becomes extremely dynamic. With such varying environment, it becomes crucial to monitor the conditions of the power cables [81, 82].

One way to monitor power cables is through the insertion of fibers in the inner conductors to measure the temperature, which is used as an indication about the circuit performance. High temperature regions, typically called hot spots in this context, can reduce power transfer capabilities of underground cables. DTSs could perfectly be used in this case to spot the position of those hot spots and help in the dynamic optimization of the power distribution with real-time monitoring. The same concept is also applicable to power cables of submarines [82, 83].

Following the same line of thought, DTSs are also proposed for fire detection in different civil infrastructures such as mines, tunnels, and industrial plants [80]. Those DTSs have already shown high performance, achieving temperature resolution of 5 °C at a distance resolution of about 5 m [83]. Clearly, the alarming thresholds are set depending on the typical ambient conditions. Basically, real-time information obtained from DTSs in concrete structures, submarine installations, oil pipelines, and aircraft wings can provide timely fault diagnostics and continuous evaluation of stress, thus help avoiding disasters and causalities [80, 84].

Another adequate application for DTS is found in the oil and gas industry, where DTSs can be used to indicate geothermal oscillations through temperature monitoring specifically for down-hole monitoring [78].

Alternatively, distributed strain sensing has been widely investigated for deployment in different applications, since its first demonstration in the late 1990s [85].

Reliable structural monitoring has been one of such applications, using distributed sensing in wide civil structures for preventive maintenance. An example deployment can be found in the Gotaalv bridge (1939), Sweden. Distributed strain sensors were used to keep track of the fatigue in the steel girders due to aging, by detecting cracks wider than 0.5 mm [83, 86].

Moreover, in addition to the fiber length instrumentation, the multi-parameter sensing capability is an even more attractive solution, where the same sensing fiber can be used for the monitoring of multiple parameters. Nevertheless, the resolution range and sensitivity requirements of a measurement differ from one application to another, rendering it impossible to find a universal instruction manual for the design of multi-parameter sensors [83].

Distributed sensors can also be used in optical fiber communications, where measurements of the optical fiber polarization spatially resolved properties are needed, and whose results may provide the guideline for the design and manufacturing of low polarization mode dispersion (PMD) fibers for speed communication systems [78].

In conclusion, it is clear that distributed sensing is highly required, when monitoring certain physical parameters as strain, temperature, and vibrations. DOFSs are highly adequate to fill the gap for such demand, due to its unique combination of low loss and sensitivity to those physical parameters. However, those advantages do not come that easy. The production of distributed OFSs requires overcoming certain challenges in order to provide a cost-effective and practical sensing system. The first challenge is instrumentation issue, such as spatial resolution or dynamic range [83]. As all optical fiber based sensors, DOFSs have to be produced such that they are able to discriminate between effects of different physical parameters, such as strain and temperature [79]. Finally, appropriate encapsulation has to be designed in order to provide intimate contact with physical structures (to achieve high sensitivity), but also to be protected from harsh environments that may reduce their efficiency [78, 80, 83].

## 2.6  Summary

The basic concepts of optical fiber sensing have been introduced in Chap. 1, along with the explanation of the different types of techniques used for OFSs. This chapter provides another step forward, by providing more details about the different types of manufactured OFSs. The production processes of the different types of OFSs are discussed. The chapter covers the different techniques of optical fiber sensing, namely intensity-based, phase-based (or interferometers), wavelength-based, polarization-based, and distributed optical fiber sensors. Different production methodologies are described for each sensor type. The challenges existing in productions processes are pointed out. Additionally, for each type of sensing technique, the most common applications are listed. Multiple applications are named per sensing type, while more than one sensing type can measure the same physical parameters. The chapter also highlights the sensitivity and performance ranges of each sensing type, while emphasizing their limitations.

This chapter presents a survey of the production methods of different types of OFSs. Readers interested in learning how to produce different types of sensors can find this knowledge, along with which sensing type is possible to use for their application and based on their setup and requirements.

# References

1. Vali, Victor, and R. W. Shorthill. "Fiber ring interferometer." *Appl. Opt* 15.5 (1976): 1099-1100.
2. Butter, Charles D., and G. B. Hocker. "Fiber optic strain gauge." *Fiber Optics Weekly Update* (1978): 245.
3. Bogue, Robert. "Fibre optic sensors: a review of today's applications."*Sensor Review* 31.4 (2011): 304-309.
4. Rajan, Ginu, ed. *Optical fiber sensors: advanced techniques and applications*. Vol. 36. CRC Press, 2015.
5. Grattan, L. S., and B. T. Meggitt, eds. *Optical Fiber Sensor Technology: Fundamentals.* Springer Science & Business Media, 2013.
6. Krohn, David A. "Intensity modulated fiber optic sensors overview."*Cambridge Symposium-Fiber/LASE'86*. International Society for Optics and Photonics, 1987.
7. Lau, Doreen, et al. "Intensity-modulated microbend fiber optic sensor for respiratory monitoring and gating during MRI." IEEE Transactions on Biomedical Engineering 60.9 (2013): 2655-2662.
8. Anwar Zawawi, Mohd, Sinead O'Keffe, and Elfed Lewis. "Intensity-modulated fiber optic sensor for health monitoring applications: a comparative review."*Sensor Review* 33.1 (2013): 57-67.
9. Xuejin, Li, et al. "Microbending optical fiber sensors and their applications."*Proceedings of the 2008 International Conference on Advanced Infocomm Technology*. ACM, 2008.
10. Lagakos, Nicholas, J. H. Cole, and Joseph A. Bucaro. "Microbend fiber-optic sensor." Applied optics 26.11 (1987): 2171-2180.
11. Gaston, Ainhoa, et al. "Evanescent wave optical-fiber sensing (temperature, relative humidity, and pH sensors)." IEEE Sensors Journal 3.6 (2003): 806-811.
12. Wang, Pengfei, et al. "High-sensitivity, evanescent field refractometric sensor based on a tapered, multimode fiber interference." Optics Letters 36.12 (2011): 2233-2235.
13. DeGrandpre, Michael D., and Lloyd W. Burgess. "Long path fiber-optic sensor for evanescent field absorbance measurements." Analytical Chemistry 60.23 (1988): 2582-2586.
14. Khijwania, S. K., and B. D. Gupta. "Maximum achievable sensitivity of the fiber optic evanescent field absorption sensor based on the U-shaped probe."*Optics Communications* 175.1 (2000): 135-137.
15. Ahmad, Mohammad, and Larry L. Hench. "Effect of taper geometries and launch angle on evanescent wave penetration depth in optical fibers."*Biosensors and Bioelectronics* 20.7 (2005): 1312-1319.
16. Grazia, Anna, Mignani Riccardo, and Falciai Leonardo Ciaccheri. "Evanescent wave absorption spectroscopy by means of bi-tapered multimode optical fibers." *Applied spectroscopy* 52.4 (1998): 546-551.
17. Monro, Tanya M., D. J. Richardson, and P. J. Bennett. "Developing holey fibers for evanescent field devices." *Electronics Letters* 35.14 (1999): 1188-1189.
18. Grillet, A., et al. "Optical fibre sensors embedded into medical textiles for monitoring of respiratory movements in MRI environment.". *Third European Workshop on Optical Fibre Sensors*. International Society for Optics and Photonics, 2007.

19. Lee, S. Thomas, et al. "A microbent fiber optic pH sensor." *Optics Communications* 205.4 (2002): 253-256.
20. Spillman Jr, W. B., et al. "A 'smart'bed for non-intrusive monitoring of patient physiological factors." *Measurement Science and Technology* 15.8 (2004): 1614.
21. Dunne, Lucy E., et al. "Design and evaluation of a wearable optical sensor for monitoring seated spinal posture." *2006 10th IEEE International Symposium on Wearable Computers.* IEEE, 2006.
22. Williams, Jonathan M., Inam Haq, and Raymond Y. Lee. "Dynamic measurement of lumbar curvature using fibre-optic sensors." *Medical engineering & physics* 32.9 (2010): 1043-1049.
23. Bilro, L., et al. "A reliable low-cost wireless and wearable gait monitoring system based on a plastic optical fibre sensor." Measurement Science and Technology 22.4 (2011): 045801.
24. Grillet, A., Kinet, D., Witt, J., Schukar, M., Krebber, K., Pirotte, F. and Depre, A. (2008), "Optical fiber sensors embedded into medical textiles for healthcare monitoring", IEEE Sensors Journal, Vol. 8 No. 7, pp. 1215-22.
25. Narbonneau, F., et al. "OFSETH: smart medical textile for continuous monitoring of respiratory motions under magnetic resonance imaging." *2009 Annual International Conference of the IEEE Engineering in Medicine and Biology Society.* IEEE, 2009.
26. Zendehnam, A., et al. "Investigation of bending loss in a single-mode optical fibre." *Pramana* 74.4 (2010): 591-603.
27. Leung, Christopher KY, et al. "Review: optical fiber sensors for civil engineering applications." Materials and Structures 48.4 (2015): 871-906.]
28. Ye, X. W., Y. H. Su, and J. P. Han. "Structural health monitoring of civil infrastructure using optical fiber sensing technology: A comprehensive review." The Scientific World Journal 2014 (2014).
29. Lee, Byeong Ha, et al. "Interferometric fiber optic sensors." Sensors 12.3 (2012): 2467-2486.
30. Tofighi, Sara, et al. "Interferometric Fiber-Optic Sensors." Optical Fiber Sensors: Advanced Techniques and Applications. Ed. Ginu Rajan. CRC Press, 2015. 37-78.
31. Turner, Dennis R. "Etch procedure for optical fibers." U.S. Patent No. 4,469,554. 4 Sep. 1984.
32. Mudhana, Gopinath, et al. "Fiber-optic probe based on a bifunctional lensed photonic crystal fiber for refractive index measurements of liquids." *IEEE Sensors Journal* 11.5 (2011): 1178-1183.
33. Choi, Hae Young, et al. "Miniature fiber-optic high temperature sensor based on a hybrid structured Fabry–Perot interferometer." *Optics letters* 33.21 (2008): 2455-2457..
34. Antunes, Paulo FC, et al. "Optical fiber microcavity strain sensors produced by the catastrophic fuse effect." IEEE Photonics Technology Letters 26.1 (2014): 78-81.
35. Domingues, M. F., et al. "Enhanced sensitivity high temperature optical fiber FPI sensor created with the catastrophic fuse effect." Microwave and Optical Technology Letters 57.4 (2015): 972-974.
36. Huang, S. C., W. W. Lin, and M. H. Chen. "Time-division multiplexing of polarization-insensitive fiber-optic Michelson interferometric sensors." *Optics letters* 20.11 (1995): 1244-1246.
37. Dandridge, Anthony D., and Alan D. Kersey. "Multiplexed interferometric fiber sensor arrays.". *Distributed and Multiplexed Fiber Optic Sensors*. International Society for Optics and Photonics, 1992.
38. Kersey, A. D., A. Dandridge, and M. A. Davis. "Low-crosstalk code-division multiplexed interferometric array." *Electronics letters* 28.4 (1992): 351-352.
39. Ryf, Roland, et al. "Space-division multiplexing over 10 km of three-mode fiber using coherent 6× 6 MIMO processing." *National Fiber Optic Engineers Conference.* Optical Society of America, 2011.
40. Houston, M. H., B. N. P. Paulsson, and L. C. Knauer. "Fiber optic sensor systems for reservoir fluids management." *Offshore Technology Conference.* Offshore Technology Conference, 2000.
41. Lim, T. K., et al. "Fiber optic acoustic hydrophone with double Mach–Zehnder interferometers for optical path length compensation." optics communications 159.4 (1999): 301-308.

42. Lin, W. W., et al. "The configuration analysis of fiber optic interferometer of hydrophones." *OCEANS'04. MTTS/IEEE TECHNO-OCEAN'04.* Vol. 2. IEEE, 2004.
43. Morris, Paul, et al. "A Fabry–Pérot fiber-optic ultrasonic hydrophone for the simultaneous measurement of temperature and acoustic pressure." The Journal of the Acoustical Society of America 125.6 (2009): 3611-3622.
44. Kuzmenko, Paul J. "F1. 5 Experimental Performance Of A Miniature Fabry-Perot Fiber Optic Hydrophone." 8th Optical Fiber Sensors Conference. 1992.
45. Li, Hong-Nan, Dong-Sheng Li, and Gang-Bing Song. "Recent applications of fiber optic sensors to health monitoring in civil engineering." Engineering structures 26.11 (2004): 1647-1657.
46. Othonos, Andreas. "Fiber bragg gratings." Review of scientific instruments 68.12 (1997): 4309-4341.
47. Sengupta, Dipankar. "Fiber Bragg Grating Sensors and Interrogation Systems." Optical Fiber Sensors: Advanced Techniques and Applications. Ed. Ginu Rajan. CRC Press, 2015. 207-256.
48. Hill, Kenneth O., and Gerald Meltz. "Fiber Bragg grating technology fundamentals and overview." Journal of lightwave technology 15.8 (1997): 1263-1276.
49. Wang, Xu-Dong, and Otto S. Wolfbeis. "Fiber-optic chemical sensors and biosensors (2008–2012)." Analytical chemistry 85.2 (2012): 487-508.
50. Kashyap, Raman. *Fiber bragg gratings.* Academic press, 1999.
51. Lam, D. K. W., and Brian K. Garside. "Characterization of single-mode optical fiber filters." *Applied Optics* 20.3 (1981): 440-445.
52. Meltz, G., W_W Morey, and W. H. Glenn. "Formation of Bragg gratings in optical fibers by a transverse holographic method." *Optics letters* 14.15 (1989): 823-825.
53. Hill, Kenneth O., et al. "Bragg gratings fabricated in monomode photosensitive optical fiber by UV exposure through a phase mask." *Applied Physics Letters* 62.10 (1993): 1035-1037.
54. Malo, B., et al. "Point-by-point fabrication of micro-Bragg gratings in photosensitive fibre using single excimer pulse refractive index modification techniques." *Electronics Letters* 29.18 (1993): 1668-1669.
55. Patrick, H., and Sarah L. Gilbert. "Growth of Bragg gratings produced by continuous-wave ultraviolet light in optical fiber." *Optics letters* 18.18 (1993): 1484-1486.
56. Simpson, A. G., et al. "Formation of type IA fibre Bragg gratings in germanosilicate optical fibre." *Electronics Letters* 40.3 (2004): 1.
57. Archambault, J-L., L. Reekie, and P. St J. Russell. "100% reflectivity Bragg reflectors produced in optical fibres by single excimer laser pulses."*Electronics Letters* 29.5 (1993): 453-455.
58. Bandyopadhyay, Somnath, et al. "Ultrahigh-temperature regenerated gratings in boron-codoped germanosilicate optical fiber using 193 nm." *Opt. Lett* 33.16 (2008): 1917-1919.
59. Albert, Jacques, Li-Yang Shao, and Christophe Caucheteur. "Tilted fiber Bragg grating sensors." *Laser & Photonics Reviews* 7.1 (2013): 83-108.
60. Byron, K. C., et al. "Fabrication of chirped Bragg gratings in photosensitive fibre." *Electronics letters* 29.18 (1993): 1659-1660.
61. Antunes, Paulo Fernando Costa, et al. "Biaxial optical accelerometer and high-angle inclinometer with temperature and cross-axis insensitivity." IEEE Sensors Journal 12.7 (2012): 2399-2406.
62. Antunes, Paulo, et al. "Dynamic monitoring and numerical modelling of communication towers with FBG based accelerometers." Journal of Constructional Steel Research 74 (2012): 58-62.
63. Antunes, Paulo, et al. "Dynamic monitoring of an elevated water reservoir with an optical biaxial accelerometer." OSA Sensors (2012): 24-28.
64. PF da Costa Antunes, Paulo Fernando, et al. "Optical fiber accelerometer system for structural dynamic monitoring." IEEE Sensors Journal 9.11 (2009): 1347-1354.
65. Butov, O. V., E. M. Dianov, and K. M. Golant. "Nitrogen-doped silica-core fibres for Bragg grating sensors operating at elevated temperatures."*Measurement science and technology* 17.5 (2006): 975.

66. Fokine, Michael. "Growth dynamics of chemical composition gratings in fluorine-doped silica optical fibers." *Optics letters* 27.22 (2002): 1974-1976.
67. J Canning, J., K. Sommer, and M. Englund. "Fibre gratings for high temperature sensor applications." *Measurement Science and Technology* 12.7 (2001): 824.
68. Correia, Sandra FH, et al. "Optical fiber relative humidity sensor based on a FBG with a diureasil coating." Sensors 12.7 (2012): 8847-8860.
69. Yeo, T. L., T. Sun, and K. T. V. Grattan. "Fibre-optic sensor technologies for humidity and moisture measurement." Sensors and Actuators A: Physical 144.2 (2008): 280-295.
70. Kronenberg, Pascal, et al. "Relative humidity sensor with optical fiber Bragg gratings." Optics letters 27.16 (2002): 1385-1387.
71. Butler, M. A., and D. S. Ginley. "Hydrogen sensing with palladium-coated optical fibers." *Journal of applied physics* 64.7 (1988): 3706-3712.
72. Woliński, T. R., P. Lesiak, and A. W. Domański. "Polarimetric optical fiber sensors of a new generation for industrial applications." *TECHNICAL SCIENCES* 56.2 (2008).
73. Hu, Wenbin, et al. "Polarization-based optical fiber sensor of steel corrosion." *SPIE Nanoscience+ Engineering*. International Society for Optics and Photonics, 2015.
74. Berry, M. V., et al. *Progress in Optics*. Ed. Emil Wolf. Vol. 50. Elsevier, 2007.
75. Ma, Jianjun, and Anand Asundi. "Structural health monitoring using a fiber optic polarimetric sensor and a fiber optic curvature sensor-static and dynamic test." *Smart materials and structures* 10.2 (2001): 181.
76. Lesiak, Piotr, et al. "All fiber optic modular sensing system for hydrostatic pressure measurements with a photonic liquid crystal fiber analyzer."*Lightguides and Their Applications III*. International Society for Optics and Photonics, 2007.
77. Ramakrishnan, Manjusha, et al. "Measurement of thermal elongation induced strain of a composite material using a polarization maintaining photonic crystal fiber sensor." Sensors and Actuators A: Physical 190 (2013): 44-51.
78. Bao, Xiaoyi, and Liang Chen. "Recent progress in distributed fiber optic sensors." Sensors 12.7 (2012): 8601-8639.
79. Galindez-Jamioy, Carlos Augusto, and José Miguel López-Higuera. "Brillouin distributed fiber sensors: an overview and applications." *Journal of Sensors* 2012 (2012).
80. Barrias, António, Joan R. Casas, and Sergi Villalba. "A Review of Distributed Optical Fiber Sensors for Civil Engineering Applications." Sensors 16.5 (2016): 748.
81. Bolognini, Gabriele, et al. "Analysis of distributed temperature sensing based on Raman scattering using OTDR coding and discrete Raman amplification."*Measurement Science and Technology* 18.10 (2007): 3211.
82. Bolognini, Gabriele, and Arthur Hartog. "Raman-based fibre sensors: Trends and applications." *Optical Fiber Technology* 19.6 (2013): 678-688.
83. Srinivasan, Balaji, and Deepa Venkitesh. "Distributed Fiber-Optic Sensors and Their Applications." Optical Fiber Sensors: Advanced Techniques and Applications. Ed. Ginu Rajan. CRC Press, 2015. 309-358.
84. Chapeleau, Xavier, et al. "Study of ballastless track structure monitoring by distributed optical fiber sensors on a real-scale mockup in laboratory."*Engineering Structures* 56 (2013): 1751-1757.
85. Horiguchi, Tsuneo, Toshio Kurashima, and Mitsuhiro Tateda. "Tensile strain dependence of Brillouin frequency shift in silica optical fibers." *IEEE Photonics Technology Letters* 1.5 (1989): 107-108.
86. Enckell, Merit, et al. "Evaluation of a large-scale bridge strain, temperature and crack monitoring with distributed fibre optic sensors." *Journal of Civil Structural Health Monitoring* 1.1-2 (2011): 37-46.

# Chapter 3
# Low Cost Silica Optical Fiber Sensors

The demonstrated advantages of optical fiber sensing systems over other sensing technologies increased the demand for these sensing mechanisms and its cost effective production [1]. According to an ElectroniCast market forecast study on Optical Fiber Sensor (OFS) consumption, the use of optical fiber sensors generated revenue of $2.79 Billion in 2015 and it is forecasted that during the 2015–2020 timeline, this value will increase on an average annual rate of 10%, reaching $4.49 billion in 2020. The extrapolated figures are shown in Fig. 3.1 [2].

It is worth highlighting here that Distributed Optical Fiber Sensor (DOFS) systems refer to the whole length of the optical fiber, which, as stated in previous chapters, can act as multi point sensing network along one single optical fiber. For the study presented by ElectroniCast, each point optical fiber sensor counts as one unit. On the other hand, the number of DOFSs is based on a complete optical fiber line, which can comprise several sensing systems. As stated by Stephen Montgomery, Director of the Fiber Optics Components group at ElectroniCast Consultants: "Since a distributed continuous optical fiber line (system) may have 100s of sensing elements in a continuous-line, it is important to note that ElectroniCast counts all of those sensing elements in a distributed continuous system as one (system) unit only. In the case of some applications, the price of the system may be several thousand dollars" [3].

Also in their market forecast, it was considered that "point sensors are often used in Distributed optical fiber sensor systems (installed at multiple-points/point-to-point); however, we count their use in the Point fiber optic sensor category and not in the continuous (non-stop) distributed sensor category," Montgomery added.

Another report by the Photonic Sensor Consortium, only on DOFS, estimated its market to be $1.1 billion in 2016, with 70% out of it being associated with the oil and gas industries [3].

In addition to the energy industries, OFSs are used in a wider range of fields, including but not limited to defense, medicine, industrial, structural, transportation, and security. Their advantageous characteristics, like light weight, small size,

© The Author(s) 2017
M.F.F. Domingues, A. Radwan, *Optical Fiber Sensors for IoT and Smart Devices*,
SpringerBriefs in Electrical and Computer Engineering,
DOI 10.1007/978-3-319-47349-9_3

**Fig. 3.1** Continuous
Distributed and Point Fiber
Optic Sensor Global
Consumption Market
Forecast (Value Basis,
$Billion) [2]

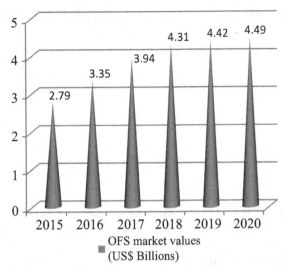

OFS market forecast values (2015-2020)

electromagnetic immunity, high resolution and accurate measures, make this type of sensing systems one of the most promising sensing solutions to monitor a diverse number of parameters such as acceleration, strain, pressure, displacement, ion concentrations, relative humidity (RH), temperature, refractive index (RI), among many others [4–8].

However, in many situations, the development, improvement and fabrication of these sensing technologies involve high financial investment and complex experimental setups [5, 6]. In order to overcome such economical drawbacks of the OFS technologies, new research achievements have been presented with low cost production and implementation techniques of OFS [1].

As for DOFS to achieve their full market potential, features, such as their robustness, must be improved. Oil and gas industry market has opened an entire new stream of business opportunities for OFS market, as they paved the way for an entire new revenue generation system for the service providers.

The advantages of optical fiber sensors over the conventional electronic sensing systems can be listed as light weight, improved reliability (no need of electronic devices at the sensor point), low cost, high bandwidth and electromagnetic immunity. Such advantages have strengthened their position in the sensing market [9].

The increasing investments in structures health monitoring (civil engineering), smart materials, the fast development in telecommunication industry, the globalization of Internet of Things (IoT) are some of the key factors instigating the growth of the global OFS market.

Nevertheless, cost production and implementation of some devices, added to some unfamiliarity with this technology, may be the main barriers to the OFS continuous growth into new fields of applications [9]; hence, as mentioned earlier, to tackle such barrier of high costs, lots of research efforts have been targeting lowering the costs of OFSs production, as well as enabling easy ways for their implementation. This chapter discusses the efforts to produce low-cost OFSs and affordable implementation methods. The first chapter of this brief has introduced the concept of optical fiber sensing and its various techniques for sensing, while the second chapter has already discussed the different production methods of the various types of OFSs. This chapter takes a step further and tackles the issue of high cost of producing OFSs. The chapter starts with a general discussion of the different opportunities available to produce a more cost-effective OFSs. The following subsections then address a more specified use-cases, including how to produce a cost-effective FBG-based OFS, or using Fiber Optic Gyroscope or Interferometric sensors to reduce production costs. The chapter also emphasizes on how distributed OFSs can offer low-cost effective solutions, especially to monitor physical parameter along larger areas, using the whole length of an optical fiber instead of point sensors. Challenges facing each technique are listed, along with some proposed solutions.

## 3.1   Cost-Effective OFS

During the past decades, several optical fiber sensing technologies were proposed and implemented. However, the needs and evaluation imposed by the field performance and the user requirements dictate that only the best solutions will pave its way to success. Only sensing techniques that perfectly adapt to the required specifications will add value to the OFS market and the sensing market in general.

Some examples of versatile and flexible solutions with high impact in the OFS market are the Fiber Bragg grating (FBG) sensors, quasi-distributed sensors, with multiplexing abilities, and reflectometry based technologies such as Brillouin, Raman and Rayleigh based sensors.

The scattering effects in silica optical fibers (i.e. Brillouin, Raman and Rayleigh) are nowadays extensively used as sensing networks, enabling the multiplexing of multiple sensing points on a remote and independent instrumented setting. When outstanding sensing performance metrics (i.e. resolution, sensitivity, and accuracy) are added to multiplexing ability, the result may just be the optimum sensing solution, with a cost effective implementation, through reducing the cost-per-measurement-point.

Nevertheless, it is important to keep in mind that the less "fit" sensing mechanisms should not be ignored or completely dismissed, since they can still be of great utility in specific small niches. As an example, Fabry-Perot Interferometers (FPI) can provide high resolution and accuracy, but they are not suitable for multiplexing network applications [10].

The following sections explore different low-cost OFS solutions and the feasibility of different technologies implementation.

## 3.2  Cost-Effective FBGs

The Fiber Bragg gratings (FBGs) implementation success and increasing demand are mostly due to the advantageous features that characterize this sensing technology, namely their tiny size (the sensing area is in the micrometers range), cost-effectiveness, insensitiveness to electromagnetic interference, which is an inherent trait that exists among all OFSs [11]. As has been previously discussed, FBG sensing mechanism is based on the monitoring of the reflected spectrum and the FBG peak position (which can be related to strain and temperature values).

Regarding this sensing technique, the higher costs may arise from the interrogation/spectrum analysis equipment, which may reach $25K, or more. The interrogation system is the device which enables the access to the Bragg wavelength and the monitoring of the shift in the peak wavelength of FBG sensor.

The majority of the FBGs in the market and the most common ones work in a wavelength window ranging from 1520 to 1590 nm, mainly due to the easily adaptable telecommunications equipment, which work in that same wavelength range [12]. Nevertheless, the interrogation equipment for such wavelength range usually increases the implementation costs of FBG sensors.

In order to overcome such financial drawback, many measures have been taken to lower such fabrication and implementation costs. To start, new lower cost interrogation mechanisms have been developed. Additionally, FBGs working in different wavelength ranges have been devised. More specifically, the use of a Bragg wavelength pick around 830–870 nm, which is typically multimode fiber (MMF) or few mode fiber (FMF), enables the use of inexpensive optical detectors (silica based), which would reduce the costs of spectrometer-based interrogators; and consequently, the investment required for the installation of FBG sensing systems. In addition to lower costs, such silica based detectors allow an increased resolution, larger bandwidth and better sampling, due to the larger number of pixels in a diode array [12].

Nevertheless, when comparing the performance of an 850 nm FBG sensor system using a few mode (FMF) 1550 nm telecom fiber (low-cost 1550 nm telecom fibers are MMF/FMF at 850 nm) to an 850 nm FBG sensor using an 850 nm SMF, it becomes evident that the presence of higher order modes (HOM) in the FMF decreases the polarization stability and diminishes the response linearity, rendering the sensing system unsuitable for high precision applications. However, the optimization of this FMF sensing system is possible, simply by coiling the FMF, and in that way restore the SMF properties of the FBG sensing system. The above proves that it is possible to have an 850 nm FBG sensor systems based on low-cost 1550 nm telecom fibers, as reported in [12].

As previously mentioned, the other way to enhance the cost effectiveness of FBG sensing systems is by creating low cost interrogation systems. Towards that end, a considerable amount of research has also been performed [13]. Among those, one of the most recent proposals presents an interrogator based on an edge filter demodulation technique, where the output voltage is linearly related to the monitored parameter [13]. The optical signal source used should have a wide bandwidth, and the reflected optical signal is scanned by different passive optical edge filters, centered at the sensor Bragg wavelength. The photodiode, on the edge filter output, returns the voltage value which is proportional to the Bragg wavelength shift. The drawback of this system is the need to associate an edge filter to each FBG sensor, limiting the number of interrogated sensors to the number of filters that can possibly fit in the interrogator; hence this solution may not be feasible in a multiplexing scenario [13].

## 3.3 Fiber Optic Gyroscope (FOG)—A Cost Effective Application of OFS

Besides the demand of low cost solutions of OFSs, there are also certain scenarios where OFSs provide a performance upgrade for similar investment costs, offering a more appealing and cost effective solution, when compared to traditional sensing systems.

One example of cost effective application of an OFS is the fiber-optic gyroscope (FOG) based on a Sagnac interferometer sensing system. A possible configuration of such solution is presented in Fig. 3.2.

This optical fiber technology was developed in the lab in the late 70s and commercially installed (in a Boeing 777) 20 years later, in the late 90s [14]. The FOG does not have any moving parts, which excludes and avoids anomalies, such as the wear out of bearings. Moreover, FOGs can be thermally isolated, and are insensitive to external vibrations and magnetic fields.

When compared to traditional gyroscopes, the benefits of the FOG are very vast [15]. Besides their reduced size, weight, volume (which are typical advantages of OFSs in general), FOGs are also favored for the fact that their response is strictly due to rotation signals—contrary to mass gyroscopes, which can also be affected by linear accelerations; making these devices highly similar or even better, than a typical rotating mass gyroscope.

Owing to such aforementioned advantages, FOGs paved their way into the inertial navigation market, making them one of the most competitive technologies in the field. Additionally, their presence is being noticed in fields like space exploration

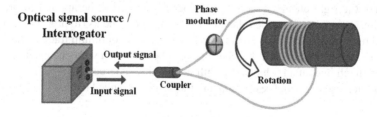

**Fig. 3.2** A FOG sensing system based on a Sagnac interferometer configuration

(need for high accuracy and immunity to linear accelerations), underwater vehicles and antenna stabilization (electromagnetic radiation immunity and small weight equipment are required). In most of the featured applications, although the implementation cost of FOGs may be in the same range as traditional gyroscope technologies, the benefits from using FOG are mirrored in its excellent performance achievements [16]; i.e. providing higher performance at the same or very similar cost.

## 3.4  Distributed OFS

As reported in the forecast presented in the beginning of this chapter, distributed OFS may be the sensing technology with the highest market penetration serving the widest profile application; hence, they receive the most attention from the optical fiber sensors market development [17].

The idea of having a plain optical fiber along the structure under surveillance (pipelines, civil structures, electric cables, etc.) and obtain a map of the temperature and/or strain along all the structure, with thermal resolutions down to the order of 1 °C, is a very appealing and cost effective solution for the continuous analysis of large areas. In fact, nowadays, the application of DOFs is becoming more common and their advantages over traditional technologies are growing more obvious to potential customers, mainly in the oil and energy industry, where thermal monitoring using DOFs proved to be profitable [17].

The growing credibility of the DOFs performance is increasing the demand for DOFs along with other OFSs to be implemented for an array of various applications, which contributes for its establishment in the sensors market.

The DOFSs have a synergy between their low loss and high sensitivity to temperature, strain and vibrations that make them distinctively positioned for the monitoring of such physical parameters [16]. Nevertheless, the road for a low cost, enduring and practical solution is still being paved, and considerable challenges still exist and must be addressed, in order to achieve the leadership in a global sensing market. Some of the foreseen challenges can be listed as follows:

- The optical fiber encapsulation: It is important when installing sensing systems to guarantee that the fiber stays physically attached (i.e. in direct contact) to the monitored structure, in order to guarantee its accurate sensitivity. On the other hand, the right encapsulation of the fiber is required to be applied in order to preserve the fiber from the influence of harsh environments (to preserve its physical reliability) [16]. This creates a dilemma and the optimum tradeoff has to be found between how hard the fiber needs to be protected with encapsulation and how much in contact it has to be with the monitored structure.
- The increase of spatial resolution and the dynamic range [16].

- The need for reliable techniques that provide distinction between different physical parameters to avoid mutual interference, which can influence the sensing system (temperature, strain and vibration) [16].

## 3.5 Interferometric Sensors

In-line optical fiber micro cavities interferometers based on single-mode fibers are solid microscopic structures, which can be cost effective and easily manufactured. Interferometric sensors have been widely used for monitoring parameters such as temperature, strain, pressure or refractive index, relative humidity, etc. [4–8, 18].

Among the existing interferometric sensing mechanisms, the Fabry Perot Interferometer (FPI) based sensors are the most widely spread ones, due its simple configurations and implementation mechanisms. Generally, the extrinsic FPI (see Chap. 1 for definition of extrinsic and intrinsic FPI) sensing devices are quite easy to manufacture, but they require a fine alignment between the cavity reflective mirrors, since their coupling efficiency may be affected (decreased) by a poor alignment [18].

Regarding the intrinsic FPI sensors, once the optical signal continuously propagates inside the optical fiber, the loss in the signal intensity is extremely low since the sensing element is a short section of fiber between two reflecting surfaces. Therefore, higher optical signal intensity will be reflected in the FPI cavity mirrors, facilitating its subsequent demodulation and analysis.

Intrinsic micro cavities can be produced through different techniques like chemical etching [19], refractive-index mismatch in the splicing area [20], internal film coating [21], laser machined and irradiated points [22], etc. However, and despite their efficiency, these intrinsic FPIs require expensive equipment and complex implementation to produce the micro cavity, which is a considerable disadvantage, when compared with other similar and more cost effective sensing mechanism [18].

Towards decreasing the production costs and overcoming such disadvantage, new low cost solutions have been proposed, regarding the production of FPI cavities for the monitoring of parameters such as hydrostatic pressure, strain, humidity, RI and high temperature [1, 4–8].

The recycling of damaged optical fiber into the production of new optical fiber FPI sensors, has proved to be a sustainable option targeting the low cost optical fiber sensing market. The method, presented by *Domingues et al.*, for the fabrication of FPI micro cavities using damaged fiber considerably reduces the experimental setup complexity and consequently, the production costs, with performance comparable to other sensors produced with higher costs [4–8]. The micro-cavities, produced using this method, were obtained from optical fiber previously damaged by the catastrophic fuse effect. The catastrophic fuse effect is a phenomenon that is usually triggered by a local heating point in the optical fiber, dirty connector or bended fiber. Such phenomenon is able to induce the optical fiber core vaporization, which permanently damages the optical fiber [23, 24]. The damaged fiber presents a sequence

**Fig. 3.3** Microscopic image of an optical fiber core damaged by the fuse effect phenomena

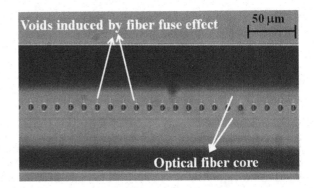

of periodic voids, as shown in Fig. 3.3, along its core, becoming unsuitable for telecommunication applications.

With this damaged optical fiber, the authors developed intrinsic FPI micro cavities by splicing a small section of the damaged fiber into a normal optical fiber, producing a micro cavity within the functional optical fiber core, with all the characteristics of an FPI sensor, reporting good performance in the monitoring of parameters such as hydrostatic pressure, high temperature, strain, refractive index, and relative humidity [1].

Following a similar approach, Zhu et al. presented a similar technique to produce a low-cost FPI, by splicing two sections of single mode optical fiber (SMF) with lateral offset between them (~62.5 μm) [25]. This type of sensor was shown to be adequate for high temperatures monitoring. Nevertheless, a pre-annealing process was required in order to obtain an acceptable linearity and repeatability. After the pre-annealing process, the FPI temperature sensitivity is comparable with the ones obtained by special optical fiber FPIs, reaching 41 nm/°C [25].

Another example of FPI micro-cavities, developed through a low cost and low complexity implementation, has been presented in multiple works [26–28]. The authors in [26, 27] presented an in-line FPI with a micro air cavity formed by splicing a single mode optical fiber (SMF) with a photonic crystal fiber (PCF). Similar to previous examples, these FPI cavities can also endure high temperatures and present relative high strain sensitivity.

Besides the splicing between the PCF and the SMF, a micro cavity can as well be generated between two SMF fibers, as presented by, Duan et al. [28]. In their method, the air micro cavity is produced simply by adjusting the splicing parameters within the splicing machine (arc splicer, Fitel S176). The produced sensors presented higher strain sensitivity, almost 150% higher than that of the results reported by comparable work previously reported. On the other hand, temperature sensitivity is only 0.848 pm/°C, which is much lower than the values of previously reported work.

The gap in the temperature and strain sensitivity is due to the micro cavity large diameter, when compared to the usually presented micro cavities. Such characteristic is extremely advantageous regarding the strain sensing applications [28].

Besides FPI sensing mechanisms, other in-line optical fiber core-cladding-mode interferometers (CCMI) have to be mentioned here. This interferometric technology has been proposed to optimize the separated arm interferometers, typically Mach Zehnder and Michelson interferometers. As the name implies, these types of sensing mechanisms operate on the interference between the cladding and the core modes. Although the method is based on an inline interferometry system, the method still requires the splitter and the combiner to couple and re-couple the optical signal between the core and the cladding modes [18].

An in-line structure in which FPI and CCMI are built (one SMF with two possible optical paths), is simple to assemble, compact, stable, with an easy alignment and therefore high coupling efficiency and with a very low implementation complexity and low fabrication cost [18].

## 3.6 Summary

Several reports in the literature and industrial sector have been confirming the widespread of the use of OFSs, across wide range of applications. The widespread of OFSs utilization has been due to their obvious advantages, discussed all over this brief. Despite the fast adoption of OFS technology, the increase in revenue and implementation would have been even more, if it were not for the high costs associated with certain functionalities or in the production process of OFSs. For OFSs to achieve their highest potential, issues with such high costs have to be addressed and solutions have to be found. This chapter discusses proposed methods to decrease the production costs of OFSs, in addition to other technologies that would help decrease associated costs of OFSs. For instance, FBGs suffer from high associated costs, due to the need for expensive interrogator in the sensing system. The chapter discusses multiple methods proposed in the field of OFS production, which mainly aim at reducing the costs associated to production of OFSs or associated to equipment required by the optical fiber sensing systems, such as cheaper interrogators. The list of discussed cost-effective solutions covers low-cost FBGs, cost-effect Fiber Optic Gyroscope and Interferometric sensors, in addition to distributed OFSs. Distributed OFSs offer a unique attractive behavior, since one distributed sensor can replace multiple point-based sensors; hence providing a very cost-effective solutions for applications requiring measurements over large areas. Basically, OFSs offer a great solution replacing traditional mechanical and electronic sensors, with similar or better performance, at lower costs, especially if the proposed solutions in this chapter are applied.

# References

1. André, Paulo, et al. "Recycling optical fibers for sensing." *SPIE Photonics Europe*. International Society for Optics and Photonics, 2016.
2. Market Research Report, "Fiber Optic Sensors Global Market Forecast & Analysis 2015-2020", *ElectroniCast*, March 18, 2016.
3. Conard Holton, "Fiber optic sensor market to see 18% growth through 2018", *LaserFocusWorld*, February 20, 2014.
4. Antunes, Paulo FC, et al. "Optical fiber microcavity strain sensors produced by the catastrophic fuse effect." *IEEE Photonics Technology Letters* 26.1 (2014): 78-81.
5. Domingues, M. F., et al. "Enhanced sensitivity high temperature optical fiber FPI sensor created with the catastrophic fuse effect." *Microwave and Optical Technology Letters* 57.4 (2015): 972-974.
6. Maria de Fátima, F. Domingues, et al. "Liquid hydrostatic pressure optical sensor based on micro-cavity produced by the catastrophic fuse effect."*IEEE Sensors Journal* 15.10 (2015): 5654-5658.
7. Domingues, M. Fátima, et al. "Cost effective refractive index sensor based on optical fiber micro cavities produced by the catastrophic fuse effect."*Measurement* 77 (2016): 265-268.
8. Alberto, N., et al. "Relative humidity sensing using micro-cavities produced by the catastrophic fuse effect." *Optical and Quantum Electronics* 48.3 (2016): 1-8.
9. Future Markets insights, "Distributed Fiber Optic Sensors Market: Global Industry Analysis and Opportunity Assessment 2015-2025", 2016.
10. Ferdinand, Pierre. "The evolution of optical fiber sensors technologies during the 35 last years and their applications in structure health monitoring."*EWSHM-7th European Workshop on Structural Health Monitoring*. 2014.
11. Kersey, Alan D., et al. "Fiber grating sensors." *Journal of lightwave technology* 15.8 (1997): 1442-1463.
12. Ganziy, Denis, Bjarke Rose, and Ole Bang. "Performance of low-cost few-mode fiber Bragg grating sensor systems: polarization sensitivity and linearity of temperature and strain response." *Applied Optics* 55.23 (2016): 6156-6161.
13. Kinet, D., et al. "Cost-effective FBG interrogation combined with cepstral-based signal processing for railway traffic monitoring.". *SPIE Photonics Europe*. International Society for Optics and Photonics, 2016.
14. Bohnert, Klaus, et al. "Highly accurate fiber-optic DC current sensor for the electrowinning industry." *IEEE Transactions on industry applications* 43.1 (2007): 180-187.
15. Lefevre, Herve C. *The fiber-optic gyroscope*. Artech house, 2014.
16. Culshaw, Brian. "Future Perspectives for Fiber-Optic Sensing." *Optical Fiber Sensors: Advanced Techniques and Applications*. Ed. Ginu Rajan. CRC Press, 2015. 521-544.
17. Hill, David John, and Magnus McEwen-King. "Distributed fiber optic sensing." U.S. Patent No. 8,923,663. 30 Dec. 2014.
18. Zhu, Tao, et al. "In-line fiber optic interferometric sensors in single-mode fibers." *Sensors* 12.8 (2012): 10430-10449.
19. Tafulo, Paula AR, et al. "Intrinsic Fabry–Pérot cavity sensor based on etched multimode graded index fiber for strain and temperature measurement." *IEEE Sensors Journal* 12.1 (2012): 8-12.
20. Rao, Yun-Jiang, et al. "In-line fiber Fabry-Perot refractive-index tip sensor based on endlessly photonic crystal fiber." *Sensors and Actuators A: Physical* 148.1 (2008): 33-38.
21. Kao, T. W., and H. F. Taylor. "High-sensitivity intrinsic fiber-optic Fabry–Perot pressure sensor." *Optics letters* 21.8 (1996): 615-617.
22. Wei, Tao, et al. "Miniaturized fiber inline Fabry-Perot interferometer fabricated with a femtosecond laser." *Optics letters* 33.6 (2008): 536-538.
23. Kashyap, R., and K. J. Blow. "Observation of catastrophic self-propelled self-focusing in optical fibers." *Electronics Letters* 24 (1988): 47-49.

24. Domingues, F., et al. "Observation of fuse effect discharge zone nonlinear velocity regime in erbium-doped fibers." *Electronics letters* 48.20 (2012): 1295-1296.
25. Zhu, Tao, et al. "Fabry–Perot optical fiber tip sensor for high temperature measurement." *Optics Communications* 283.19 (2010): 3683-3685.
26. Villatoro, Joel, et al. "Photonic-crystal-fiber-enabled micro-Fabry–Perot interferometer." *Optics letters* 34.16 (2009): 2441-2443.
27. Deng, Ming, et al. "PCF-based Fabry–Pérot interferometric sensor for strain measurement at high temperatures." *IEEE Photonics Technology Letters* 23.11 (2011): 700-702.
28. Duan, De-Wen, et al. "Microbubble based fiber-optic Fabry–Perot interferometer formed by fusion splicing single-mode fibers for strain measurement." *Applied optics* 51.8 (2012): 1033-1036.

# Chapter 4
# Polymer Optical Fiber Sensors

In previous chapters, different optical sensing mechanisms have been discussed. The characteristics of the different types of OFSs have been presented and compared. Although different types and characteristics have been shown, all those sensing techniques were fabricated using silica optical fibers. In this chapter, an alternative line of optical fiber sensors is introduced. Those sensors are fabricated using polymer (in other words plastic). The chapter discusses this type of polymer-based OFSs, elaborating on the different sensing techniques and wide range of applications. More importantly, this type of OFSs is compared to the more commonly used Silica based OFSs. The chapter provides an introduction to polymer optical fibers and their use in sensing. The characteristics of POFs are then discussed, highlighting their advantages and limitations. The different techniques used for sensing in POFs are then presented, with demonstration of their most potential applications and the motivation behind using this alternative type of OFSs.

## 4.1 Introduction to POF

Till recently, optical fiber sensing market was dominated by silica OFSs, due to their ease of adaptation to the equipment and devices already in use in the telecommunication field.

Nowadays, Polymer optical fiber, also known as Plastic optical fiber (POF), is growing as an economical alternative to silica optical fiber and its use for sensing purposes has seen a considerable increase in the last decade. Their unique properties of sensing parameters such as strain, refractive index, temperature, pressure, relative humidity, together with the low cost associated to this type of fibers and their appealing characteristics (development of single mode fibers, microstructured fibers, etc.) gave this technology a promising future in the optical fiber sensors market [1].

© The Author(s) 2017
M.F.F. Domingues, A. Radwan, *Optical Fiber Sensors for IoT and Smart Devices*,
SpringerBriefs in Electrical and Computer Engineering,
DOI 10.1007/978-3-319-47349-9_4

Plastic optical fibers (POFs) have been under research for the last 3 decades. It was a slow progress mainly due to its high transmission loss at that time, and the lack of a market suitable for this technology. However, with the improvement in the refractive index characteristics of the optical fiber, namely the production of graded index fibers and the consequent decrease of in transmission loss, POF has returned to the spotlight, as it opens new fields of applications and provides opportunity for the development of cost effective sensing mechanisms [2, 3].

In the field of telecommunications, POF offers the advantages inherent to optical fibers, which include reduced size, flexibility, cost effective, immunity to interference, etc. Although its application is mainly concentrated in short haul communications (due its attenuation levels for long distances), they offer a considerable support for silica optical fiber in communication links.

On the other hand, its application in the sensing field has a wide range of applications. Its vast application field ranges from biomedicine [4, 5], to structural health monitoring and liquid level detection [6, 7]. POF sensing technology can be applied in monitoring of strain, temperature, pressure, turbidity, density, position, refractive index, just to name a few examples [2, 8]. It is highly expected that, with the evolution and rise of new and optimized types of POFs, its action scope will widen even more [2].

POFs characteristics—high elastic strain limits, bending flexibility, high fracture toughness and potential negative thermal sensitivities—give them major advantages in diverse sensing applications. Its unique advantages include the high ductility of polymer materials and the low stiffness added to its reduced size and weight. Such features allow POF sensors to be implemented in environments with high strain levels and where precise devices are required [8].

## 4.2  POF Characteristics

Polymer optical Fibers have a geometry similar to the one presented for silica optical fibers. They are composed of a core, a cladding and often also a jacket (for more resistance), as illustrated in Fig. 4.1. Despite the high similarity shown in the figure, nevertheless, the core of POF fibers has a diameter considerably larger (1 mm) than that of Silica fibers (usually ~10 μm).

Nowadays, for the production of POFs, a variety of polymers are used, among which the polymethyl-methacrylate (PMMA), polycarbonate (PC), polystyrene (PS) and amorphous fluorinated polymer (CYTOP) are the most common [9].

Multimode POFs are available with a wide variety of diameters, larger than the typical characteristic of the single-mode silica optical fibers. Also, they are available with diverse cross-sectional index distributions, including gradient index configurations and step index [10]. POFs are most commonly available as multimode fibers, due to their large diameter, which allows the propagation of several modes. One of the advantages of large core diameters of POFS is the simplicity and ease during cleaving and connecting. This easiness in handling, although only applicable to

**Fig. 4.1** Polymer optical
fiber basic structure

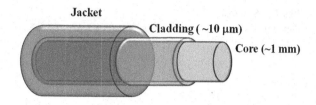

multimode POF (single mode POF requires some extra managing care), helps to spread POF sensors as an effortless solution (when compared to silica OFS), economical and implementation wise [2].

Depending on the polymer used in POF production, the properties of the produced fiber change, and so does their response to strain and temperature. When compared to silica optical fibers, their sensitivity is generally higher, showing also some negative sensitivity to temperature. POFs also provide a considerable high tolerance to bending. They are biocompatible, hence eliminating the need for the use of other special coatings, rendering them ideal for eHealth application [8].

As any other technology, POFs certainly have some limitations. In principal, they experience low thermal threshold. Other limitations are the viscoelastic properties of the polymer used and the difficulty in coupling process for different core diameter fibers (cross-sections fluctuations). In distributed sensing scenarios, specifically for long distances, the transmission losses and attenuations inherent to POF sensors may pose a considerable drawback [10].

The attenuation effect can be resulting from intrinsic or extrinsic origins. Intrinsic origins are basically due to the material properties, such as absorption, scattering, etc. Extrinsic origins occur due to material deformations, or microbending, etc. The intrinsic attenuation in POFs is of some considerable magnitude, when compared to silica optical fibers. Moreover, for wavelengths higher than 700 nm, the attenuation in POFs tends to increase, whereas in silica fibers the attenuation in contrast decreases [11]. Therefore, the wavelength windows, in which POF sensors work, are considerably different than the ones used in silica optical fiber sensors. Multimode POFs usually operate in wavelengths in the range of 400–700 nm (low intrinsic attenuation), ~850 nm (telecom applications) and above 1300 nm for the CYTOP fibers. Alternatively, silica optical fibers operate in the window of 1300–1600 nm [11].

The attenuation in POFs is a critical issue that could easily be handled during the design of POF-based sensors. As a consequence, two factors should be carefully analyzed and adjusted: the length of the sensor and the operating wavelength window.

To overcome some of the drawbacks imposed by the polymer attenuation, graded index profiles can be used in order to lower the intermodal dispersion. The implementation of graded index POFs (GIPOF) lowers the intermodal dispersion and refines the resolution of the signal transmitted over longer distances. Additionally, such implementation improves the coupling efficiency and reduces bending losses. All effects are considered advantageous features, when targeting sensing applications [11].

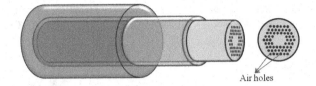

**Fig. 4.2** Schematic diagram of a microstructured POF

Air holes

Until recently and most usually, the increase of the POF core refraction index is achieved through doping; nevertheless, there have already been some reports for a dopant free fabrication processes of GIPOF [12, 13].

An alternative solution, to overcome the attenuation drawbacks of single POF, is the use of microstructured polymer optical fibers (mPOFs), in which the fiber cross section is composed of air holes, as illustrated in Fig. 4.2.

These types of fiber offer key advantages, not only over conventional POFs, but also over traditional silica optical fibers. Their production is economically appealing and the polymers malleability opened a wide range of potential applications, both in the field of telecommunications and for sensing applications. In the case of mPOFs, the attenuation is reduced by decreasing the intrinsic losses of the fiber, since the existence of air holes implies a decrease in the material responsible for the signal absorptions (intrinsic attenuation) [14].

Regarding its physical properties, POF generally presents a yield strain of 1–6%, as compared to 1–5% for silica optical fibers. The initial elastic modulus of POFs is in the interval of 1–5 GPa [15, 16]. The thermal sensitivity, when considering a bulk PolyMethylMethAcrylat (PMMA) POF, is significantly larger (~57%) than the values recorded for silica fibers, and they also provide negative values for temperature sensitivity. Such negative sensitivity to temperature offers new possibilities in fiber sensing applications, regarding the strain and temperature compensation [16].

In addition to strain and temperature sensitivity, POFs have another appealing property, when it comes to humidity. In contrast to silica optical fiber (which requires special coatings to absorb humidity and induce strain), POFs are intrinsically sensitive to humidity. An increase in the relative humidity (RH) directly induces an expansion of the PMMA and consequently a change in the refractive index [17]. This inherent characteristic of absorbing humidity in POFs is an optimum trait for the production of chemical sensors, since POFs may be able to infiltrate different chemical substances [17–19].

## 4.3   POF Sensing Mechanisms and Applications

As mentioned in previous sections, POF characteristics render them useful for sensing implementations. Their versatility and multiple features give the user the option to choose from low cost and easy handling multimode POFs to any special POF fibers with all possible enhancement of the physical attributes mentioned above, since they can be manufactured from diverse range of polymer materials.

The following sub-sections explore some phenomena used for POF sensing applications and describe some of the most common type of sensing technologies in POF, namely intensity-based, interferometry, distributed sensing, FBGs in POF and microstructured POF sensors.

## 4.3.1  Intensity

Polymeric optical fiber sensors, based on optical losses, are the most cost effective and non-complex type of sensing mechanisms, since their implementation is easy and straight forward.

These sensors owe their low cost feature to the low cost of the multimode POFs, along with the optical sources and detectors. A simple sensing system can be composed of an optical source (laser or a light emitting diode (LED)), the POF section and a photodiode to convert the light into current/voltage values. The optical intensity modulation can be achieved by decoupling two fibers; thus changing the optical transmission path from one of the POF away from the acceptance cone. Additionally, doing so changes the amount of input optical power coupled into the POF [20]. Another way, to achieve the optical intensity modulation, is by creating losses along the length of the optical fiber, by inducing a bend or grooves in the POF [8, 21, 22]. These are just a couple of examples. A schematic example is presented in Fig. 4.3.

The simplest and earliest configuration of a POF intensity sensor is based on the transmission losses induced by bended section of the polymeric optical fiber. Also, transmission losses can be achieved by lateral polish or cut in along the fiber length [7, 21].

Based on this principle, some sensing mechanisms were successfully presented. For instance, a simple implementation takes advantage of the high strain sensitivity of POF, by inserting a polymeric fiber in a section epoxy composite laminates, making it possible to detect defects and cracks in the epoxy structure. The configuration implemented only needs a LED as the optical source and a photodiode to monitor the output intensity at the POF end connection. The fact that the polymeric fiber has a high strain threshold enables the sensing mechanism to survive the epoxy degradation and to accurately retrieve the local cracks gradient growth [22, 23]. The same experimental configuration can be used to monitor strain and the bending evolution in loaded structures. The flexural and axial loads, induced in the structure, modulate light intensity in the sensing POF, which can be detected by a photodiode [24].

Apart from the simplistic (but effective) implementation described, the sensing mechanism can still be optimized to fit different application scenarios and for the measure of other parameters besides strain. The thermal optical characteristics of polymeric materials allow the design of temperature sensors, with linear response at a fixed bend radius of the POF [25]. This response is due to the fact that a variation in the temperature induces a shift of the POF refractive index and therefore a change in the optical loss intensity through the bended section of the fiber. Using this

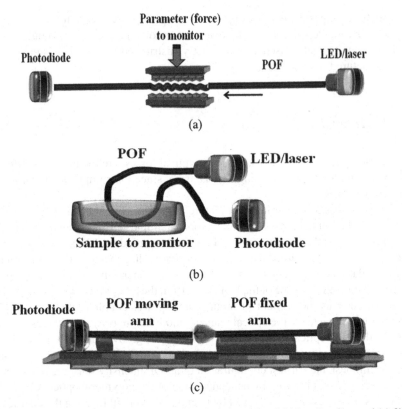

**Fig. 4.3** Intensity based POF sensing mechanism based on (**a**) and (**b**) bending fiber and (**c**) fiber coupling

method, it is possible to monitor the temperature shifts, by observing the optical loss variation [25, 26].

Also environmental refractive index (RI) changes can be monitored with POF bend based sensors in a U-Shape configuration. The optical loss at the bending induces an evanescent wave, which can be modulated by the changes in the environmental (mostly fluids) RI, in a range of 1.33–1.45 and with sensitivity of nearly 800%/RIU [27]. For RI monitoring accuracy and repeatability, the bending radius in the POF has to be controlled. This configuration, based on the RI changes with the concentration levels, has also been used to measure parameters like water salinity or TNT concentrations in alcohol [1].

The sensitivity optimization of intensity-based POF sensors is achieved by increasing the transmission power loss. That boost in the power loss can be accomplished by making groves in a cross section of the fiber. This technique has been used in the production of sensors for biomedical application [5, 28, 29], liquid level and structural health monitoring [7], among others, offering a reliable, accurate and low cost solution for those implementations.

The devices, described above, are based on the interaction of one fiber with the surrounding environment, being therefore intrinsic solutions. In addition to those technologies, other sensing systems were developed, based on an extrinsic sensing configuration. In such configuration, the sensing principle relies on the transfer of the light signal, and therefore the coupling efficiency, between two POFs (one is fixed into an inertial arm and the second moves according to the structure oscillation); or between one POF and a reflective surface, taking advantage of the scattering phenomenon along with its coupling efficiency.

Devices, using such principles, have been developed and implemented for biomedical applications (cardio respiratory monitoring) in addition to structure health applications [4, 6, 20].

### 4.3.2  Interferometry

The production of POF interferometric sensors is quite straightforward and considerably low-cost; being one of the reasons this technology has been successfully implemented for the measuring of a wide range of low frequency strain, surpassing the performance of silica based sensors in sensitivity and failure threshold [30–34]. In Fig. 4.4, the basic configuration of an interferometric sensor is presented. The interferometry in POF sensing devices can be incoherent or coherent, according to the type of fiber and the design of the sensing setup mechanism.

The time of flight interferometer is an example of an incoherent interferometer configuration, which can use multimode POF, taking advantage of its low cost and the fact that it does not need phase measurements [33]. This principle has been used to monitor vibrations in moving structures, as well as for biomedical applications [32, 35].

The POF sensors, using a coherent interferometry, require a rigorous control of the modes propagating in the fiber. Consequently, the use of single mode POF in their production is more frequent, which can additionally increases the costs of the sensing devices considerably. Nevertheless, they bring the advantages of high resolution and precision, which sometimes prove cost effective for certain sensing requirements [33].

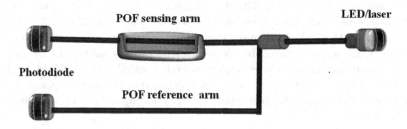

**Fig. 4.4** Schematic of an interferometric POF based sensor

### 4.3.3   Distributed Sensing

POF sensing characteristics have also been used for distributed sensing applications. Nevertheless, working strain range is quite different than the one visible in silica fibers, since the backscattering signal observed in POFs has considerably weaker sensitivity to small strains, but substantial sensitivity for larger strains [36]. Owing to those characteristics, its application over wide structures with large strain variations is ideal. Moreover, the fact that POFs can endure considerable amount of deformation (contrary to silica fibers) enables them to be deployed in different fields of application, where silica cannot perform due to endurance limitation. The deformation induced in POF devices modifies the loss properties and scattering levels in the fiber, reaching values that can be comfortably monitored [1].

POF distributed sensing mechanisms have been used for structural health monitoring and biomedical applications [37, 38]. For a respiration monitoring device, POF is suitable for embedding into a textile composite and the elongation of the POF induced by the thoracic breathing movements can be monitored by an OTDR [38].

### 4.3.4   FBGs in POF

FBG based OFSs are probably the most widely implemented type of silica optical sensors. Its wide dissemination is due to its multiplexing abilities and wavelength encode, basically meaning that the sensor is not influenced by laser fluctuations or coupling misalignments. The success of this sensing technology in silica fibers boosted the research into the development of FBG based sensors using POFs.

The first reported inscription of an FBG in a POF fiber was performed by Xiong et al., in 1999 [39]. The authors inscribe the grating in a single mode photosensitive polymeric fiber by exposing it to UV light [39].

Due to the POF high strain sensitivity and failure threshold, the FBG sensors can present a large wavelength shift in tensile strain loadings (~52 nm, considerably higher than the observed for silica FBGs) [40]. The POF FBG sensing feedback to temperature is also verified, although the thermal resistance of POF fibers is considerably lower than that of silica; hence limiting FBG-based POF sensors for lower temperature applications [41]. The challenge regarding this type of sensors is the recording of the Bragg grating into multimode POF, since not only the production costs are considerably lower than single mode fibers, but also because its implementation is much less complex. The photosensitivity of the multimode POF can be increased with dopant substances, such as benzyl dimethyl ketal (BDK) [42]. Also, in some situations, the FBG can be erased if exposed to elevated temperatures for extended periods of time. Such drawbacks in FBG sensing still need to be addressed and resolved.

Although the use of POF considerably reduces the sensor cost and increases the coupling efficiency, some extra care should be taken (coupling wise) in order to maintain the reflected spectrum and the optical power coupled in each mode, so it remains stable and it does not affect the repeatability of the measures [42].

The dominant physical mechanisms behind the Bragg grating production in polymeric fibers is the photo polymerization [8], which is different than the process used in silica fibers. Nevertheless, similar to the inscription of Bragg gratings in silica fibers, the depth of the grating also increases with extended exposure time, till it reaches a certain threshold, beyond which the fiber material may be damaged [43].

Although some great advances have been made for FBG in POFs, some key issues must yet be addressed and are currently being investigated, in order to address its pitfalls and make this technology as low cost and easy to implement as other POF sensing technologies.

### 4.3.5   Microstructure POF Sensors

As previously stated, single-mode POFs offer high-resolution sensing, but with several challenges regarding their production methods and coupling efficiency, being only suitable for short sensing lengths. A promising solution to surpass such pitfalls is the use of microstructured polymer optical fibers (mPOFs) for the production of sensors. This type of fibers has flexible hole patterns, low melting point and is considerably easy to handle [44].

The large air gap created by the microstructure in the cross section of an mPOF helps reduce the intrinsic attenuation and favors the confinement effect of the optical signal, enabling the propagation of a single mode signal for a wide wavelength range [44].

The unique microstructured cross section of mPOFs can be applied for the monitoring of force or temperature by inducing a selective liquid infiltration (one particular cladding air hole) into the fiber microstructure [45].

The fact, that the individual holes in the fiber have extremely small diameters and can be filled with fluids, is another high advantage of the mPOF, which enables them to be used for biomedical applications allowing small size samples to be analyzed with extremely high sensitivity. Additionally, the sensor can be chemically selectively reactive, by coating the microcavities with the appropriate chemicals [46, 47].

Microstructured POFs present an optimum advantage over solid core single mode POFs, since they allow the writing of highly tunable FBG sensors [10].

Nevertheless, due to the multiple lateral interfaces of mPOF to side illumination, the inscription of FBGs becomes quite more challenging. To overcome such challenging procedures, alternative innovative methods have been investigated, such as using PMMA (polymethyl methacrylate) and Topas (cyclic olefin copolymer) mPOF fiber [48, 49]. Topas has no monomers, a high glass transition ($T_g \sim 135\ ^\circ C$) and a very low moisture absorption rate (about 100 times lower than PMMA), which provides favorable conditions for the optical fiber drawing and does not induce apparent thermal strain in the sensing area [10]. FBGs written in Topas mPOFs are suitable for monitoring large strain ranges and temperatures up to approximately 110 °C [49].

It is clear now that POFs offer an alternative opportunity for OFSs, which is cheap, but they still need further investigations to overcome their drawbacks, if they would ever achieve their foreseen potential.

## 4.4  Summary

This brief represents a full review of the concept of optical fiber sensing; therefore, the brief has to cover every aspect of optical fiber sensing. The previous chapters have initially presented the general idea of sensing based on optical fiber, then went through different types of sensing methodologies, listing the main technological basics of each, along with their production methods, advantages and inherent limitations. Moreover, ideas to reduce the production costs of OFSs and the overall sensing systems have been discussed. However, all previously mentioned sensing technologies have been built using Silica optical fibers. To provide a complete view of optical fiber sensing, this chapter provided another alternative for the production of OFSs, using Polymer (or plastic) optical fibers (POFs). The chapter provided an overview of optical fibers made of polymer. The main characteristics of POFs have been presented, while always comparing to those of Silica optical fibers. This comparison would allow readers to potentially choose the optical fiber, which is best matched to the requirements of the applications and the environment. The details of the different production methods of POFs have been overviewed, coving intensity-based, interferometric, FBG-based and microstructure POF sensors. To provide a complete vision of POFs, distribution features of POF based sensors are additionally discussed. Fields, where POF-based sensors outperform silica-based ones, have been emphasized, providing a useful guide for users to decide when to use POF-based sensors. Basically, POFs provide a cost-effective alternative to silica based sensing technologies, but some challenges and issues (which has been illustrated in the chapter) have to be carefully addressed, while taking into consideration the limitations and range of applicability of POFs.

## References

1. Peters, Kara. "Polymer optical fiber sensors—a review." *Smart materials and structures* 20.1 (2010): 013002.
2. Ohtsuka, Yasuji, Eisuke Nihei, and Yasuhiro Koike. "Graded-index optical fibers of methyl methacrylate-vinyl benzoate copolymer with low loss and high bandwidth." *Applied physics letters* 57.2 (1990): 120-122.

3. Zubia, Joseba, and Jon Arrue. "Plastic optical fibers: An introduction to their technological processes and applications." *Optical Fiber Technology* 7.2 (2001): 101-140.
4. Leitão, Cátia Sofia Jorge, et al. "Plastic optical fiber sensor for noninvasive arterial pulse waveform monitoring." *IEEE Sensors Journal* 15.1 (2015): 14-18.
5. Bilro, Lúcia, et al. "Gait monitoring with a wearable plastic optical sensor." *Sensors, 2008 IEEE*. IEEE, 2008.
6. Costa Antunes, Paulo, et al. "Dynamic structural health monitoring of a civil engineering structure with a POF accelerometer." *Sensor Review* 34.1 (2014): 36-41.
7. Antunes, Paulo, et al. "Liquid level gauge based in plastic optical fiber." *Measurement* 66 (2015): 238-243.
8. Bilro, Lúcia, et al. "Optical sensors based on plastic fibers." *Sensors 12.9* (2012): 12184-12207.
9. Ziemann, Olaf, et al. "POF handbook." *Springer* (2008).
10. Peters, Kara. "4 Polymer Optical Fiber Sensors." Optical Fiber Sensors: Advanced Techniques and Applications 36 (2015): 79.
11. Ziemann, Olaf, et al. *POF-polymer optical fibers for data communication*. Springer Science & Business Media, 2013.
12. Jasim, Ali Abdulhadi, et al. "Refractive index and strain sensing using inline Mach–Zehnder interferometer comprising perfluorinated graded-index plastic optical fiber." *Sensors and Actuators A: Physical 219* (2014): 94-99.
13. Bunge, Christian-A., et al. "Dopant-free fabrication process for graded-index polymer optical fiber solely based on temperature treatment." *2015 17th International Conference on Transparent Optical Networks (ICTON)*. IEEE, 2015.
14. Van Eijkelenborg, Martijn A., et al. "Microstructured polymer optical fibre. "*Optics express* 9.7 (2001): 319-327.
15. Yang, D. X., et al. "Structural and mechanical properties of polymeric optical fiber." *Materials Science and Engineering: A* 364.1 (2004): 256-259.
16. Silva-López, Manuel, et al. "Strain and temperature sensitivity of a single-mode polymer optical fiber." *Optics Letters* 30.23 (2005): 3129-3131.
17. Zhang, Wei, David J. Webb, and G-D. Peng. "Investigation into time response of polymer fiber Bragg grating based humidity sensors." *Journal of lightwave technology* 30.8 (2012): 1090-1096.
18. Zhang, Wei, and David J. Webb. "Humidity responsivity of poly (methyl methacrylate)-based optical fiber Bragg grating sensors." *Optics letters* 39.10 (2014): 3026-3029.
19. Batumalay, Malathy, et al. "A study of relative humidity fiber-optic sensors." *IEEE Sensors Journal* 15.3 (2015): 1945-1950.
20. Antunes, Paulo Fernando Costa, Humberto Varum, and Paulo S. Andre. "Intensity-encoded polymer optical fiber accelerometer." *IEEE Sensors Journal* 13.5 (2013): 1716-1720.
21. André, P. S., et al. "Monitoring of the concrete curing process using plastic optical fibers." *Measurement* 45.3 (2012): 556-560.
22. Takeda, Nobuo. "Characterization of microscopic damage in composite laminates and real-time monitoring by embedded optical fiber sensors." *International Journal of Fatigue* 24.2 (2002): 281-289.
23. Chen, Tao, et al. "Crack detection and monitoring in viscoelastic solids using polymer optical fiber sensors." *Review of Scientific Instruments* 87.3 (2016): 035005.
24. Kuang, K. S. C., et al. "Plastic optical fibre sensors for structural health monitoring: a review of recent progress." *Journal of Sensors* 2009 (2009).
25. Moraleda, Alberto Tapetado, et al. "A temperature sensor based on a polymer optical fiber macro-bend." *Sensors* 13.10 (2013): 13076-13089.
26. Teng, Chuan-xin, Ning Jing, and Jie Zheng. "The influence of temperature to a refractive index sensor based on a macro-bending tapered plastic optical fiber." *Optical Fiber Technology* 31 (2016): 32-35.

27. Teng, Chuanxin, et al. "Investigation of a Macro-Bending Tapered Plastic Optical Fiber for Refractive Index Sensing." *IEEE Sensors Journal* 16.20 (2016): 7521-7525.
28. Stupar, Dragan Z., et al. "Wearable low-cost system for human joint movements monitoring based on fiber-optic curvature sensor." *IEEE Sensors Journal* 12.12 (2012): 3424-3431.
29. Anwar Zawawi, Mohd, Sinead O'Keffe, and Elfed Lewis. "Intensity-modulated fiber optic sensor for health monitoring applications: a comparative review." *Sensor Review* 33.1 (2013): 57-67.
30. Silva-López, Manuel, et al. "Strain and temperature sensitivity of a single-mode polymer optical fiber." *Optics Letters* 30.23 (2005): 3129-3131.
31. Kiesel, Sharon, et al. "Large deformation in-fiber polymer optical fiber sensor." *IEEE Photonics Technology Letters* 20.6 (2008): 416-418.
32. Gallego, Daniel, and Horacio Lamela. "High-sensitivity ultrasound interferometric single-mode polymer optical fiber sensors for biomedical applications." *Optics letters* 34.12 (2009): 1807-1809.
33. Kiesel, Sharon, et al. "Polymer optical fiber sensors for the civil infrastructure." *Smart Structures and Materials*. International Society for Optics and Photonics, 2006.
34. Jiang, Guoliang, et al. "Oscillator interrogated time-of-flight optical fiber interferometer for global strain measurements." *Sensors and Actuators A: Physical* 135.2 (2007): 443-450.
35. Durana, Gaizka, et al. "Use of a novel fiber optical strain sensor for monitoring the vertical deflection of an aircraft flap." *IEEE sensors journal* 9.10 (2009): 1219-1225.
36. Liehr, Sascha, et al. "Polymer optical fiber sensors for distributed strain measurement and application in structural health monitoring." *IEEE Sensors Journal* 9.11 (2009): 1330-1338.
37. Leung, Christopher KY, et al. "Review: optical fiber sensors for civil engineering applications." *Materials and Structures* 48.4 (2015): 871-906.
38. Witt, Jens, et al. "Medical textiles with embedded fiber optic sensors for monitoring of respiratory movement." *IEEE Sensors Journal* 12.1 (2012): 246-254.
39. Peng, G. D., Z. Xiong, and P. L. Chu. "Photosensitivity and gratings in dye-doped polymer optical fibers." *Optical Fiber Technology* 5.2 (1999): 242-251.
40. Liu, H. Y., H. B. Liu, and G. D. Peng. "Tensile strain characterization of polymer optical fibre Bragg gratings." *Optics Communications* 251.1 (2005): 37-43.
41. Liu, H. Y., G. D. Peng, and P. L. Chu. "Thermal tuning of polymer optical fiber Bragg gratings." *IEEE Photonics Technology Letters* 13.8 (2001): 824-826.
42. Luo, Yanhua, et al. "Analysis of multimode POF gratings in stress and strain sensing applications." *Optical Fiber Technology* 17.3 (2011): 201-209.
43. Liu, H. B., et al. "Novel growth behaviors of fiber Bragg gratings in polymer optical fiber under UV irradiation with low power." *IEEE Photonics Technology Letters* 16.1 (2004): 159-161.
44. Emiliyanov, Grigoriy, et al. "Localized biosensing with Topas microstructured polymer optical fiber." *Optics Letters* 32.5 (2007): 460-462.
45. Yang, Chunxue, et al. "Selectively liquid-infiltrated microstructured optical fiber for simultaneous temperature and force measurement." *IEEE Photonics Journal* 6.2 (2014): 1-8.
46. Peng, Lirong, et al. "Gaseous ammonia fluorescence probe based on cellulose acetate modified microstructured optical fiber." *Optics Communications* 284.19 (2011): 4810-4814.
47. Wolfbeis, Otto S., and Hermann E. Posch. "Fibre-optic fluorescing sensor for ammonia." *Analytica Chimica Acta* 185 (1986): 321-327.
48. Johnson, Ian P., Kyriacos Kalli, and David J. Webb. "827 nm Bragg grating sensor in multi-mode microstructured polymer optical fibre." *Electronics letters* 46.17 (2010): 1.
49. Markos, Christos, et al. "High-T g TOPAS microstructured polymer optical fiber for fiber Bragg grating strain sensing at 110 degrees." *Optics express* 21.4 (2013): 4758-4765.

# Chapter 5
# Optical Fiber Sensors in IoT

In previous chapters, the concept of optical fiber sensing has been thoroughly discussed. The multiple different techniques of optical fiber sensing have been presented. Different types have been compared. Silica fibers, as well as plastic (also known as polymer) fibers, have been explained. The different physical parameters, that could be measured/monitored using OFSs, have been listed, along with best options to use and limitations, based on the details of the application and required precision. Methods to reduce the costs of fabricating and implementing different OFSs have been explained, in order to increase the feasibility and applicability of optical fiber based sensing, specifically for the deployment in the field of Internet of Things (IoT), which is the buzz word of today's technology. Having paved the way, this chapter discusses the use of OFSs in IoT. The current chapter introduces the concept of IoT, highlighting the factors and motivation pushing towards a more connected world; hence the so-called Internet of Things (IoT) and in the more extreme scenario the Internet of Everything (IoE). To begin, the chapter explains the term "Internet of Things" in the first subsection, emphasizing the need for such concept, highlighting the enablers of IoT and their added value. The chapter then moves forward to highlight how OFSs can be integrated in IoT applications, to provide a more added value at lower costs. After a brief introduction of IoT, each subsection tackles a different category of IoT applications, where OFSs are seen to play a role in the very near future. It is worth mentioning here that the current chapter is not intended to provide a complete view of IoT and should not be treated as one. The chapter is alternatively dedicated towards discussing the use of OFSs within IoT. Readers, interested in learning more about IoT, are advised to read publications devoted to IoT, such as surveys by Al-Fuqaha et al. [1], or Atzori and Iera [2].

© The Author(s) 2017
M.F.F. Domingues, A. Radwan, *Optical Fiber Sensors for IoT and Smart Devices*,
SpringerBriefs in Electrical and Computer Engineering,
DOI 10.1007/978-3-319-47349-9_5

## 5.1  Introduction to IoT

The term "Internet of Things" refers to technologies that enable the networking of "Things", allowing different devices and items to sense each other and their surrounding environment, communicate among themselves, and automatically react according to situations they sense, without human interference [1, 3]. The term "Things" in the phrase "Internet of Things" refers to any object that exists in the real world, where a sensor and a communicating device (i.e. a transceiver to send and receive information) are embedded, to enable those devices to sense certain parameters and to communicate with the outside world and other smart devices (i.e. Things). Such communication usually occurs wirelessly; hence allowing the free mobility and ubiquity of those smart devices. The "Things" (smart devices) can basically be anything, as for instance vehicles, machines, buildings, people, animals, goods or even the environment surrounding those devices [2, 3]. IoT applications widely vary and include different scenarios, which are considered unlimited and cannot really be fully listed. Experts even usually assume that there are still many more applications; some are not yet known and will be discovered/introduced in the near or far future. Some more common examples include smart home (smart housing), eHealth, intelligent transport system (ITS), to name a few [1]. Within the concept of smart housing, users are able to monitor what is happening in their own homes, while away. Smart housing would also automatically adjust the house temperature according to context. For instance, in winter, heaters could be switched off, when no one is home to save energy and reduce electricity bills, while switching them back on just in time to have adequate warm temperature when home owners arrive. Smart housing can even be programmed to control coffee machines, so coffee is ready when owners wake up. Washing machines may also be controlled within the concept of smart housing, so they are only switched on when power costs are at lower rates in smart-metering scenario. Moving outside of our houses, intelligent transport system (ITS) would come in play, where our cars would be able to communicate with traffic lights, other cars and other types of transport vehicles (i.e. trains, metros, busses), in order to provide a safer and faster way of transportation. Moreover, self-driving cars are under experimentation and field trials in multiple venues, and it is just a matter of time before they are a reality and not just a technology of the future. A final example here is eHealth, which is already part of our life nowadays. Wearable and other type of health monitoring devices monitor health conditions of patients, elder citizens and even normal healthy person, for alerting users and their doctors of critical health conditions, in addition to tracking vital statistics for healthy life. eHealth can also enable remote diagnosis of some illnesses and certain symptoms, through videoconferencing among other technologies. Basically, IoT is becoming big part of our daily life at the moment, and soon enough will be present in every aspect of any person's life.

To put things in perspective, one way of categorizing IoT would be by dividing into six key, interconnected areas, namely: Smart Materials, Wearable devices, eHealth, Smart Cities, Smart Mobility, and Smart Houses [3]. This is one vision that

**Fig. 5.1** One vision of how IoT can be divided

is illustrated in Fig. 5.1. Other categorization has been introduced, such as the division into different vertical markets, namely smart home, vehicles, school, market, industry, transportation, healthcare and agriculture, as presented in [1].

IoT is envisioned to connect billions of different types of devices using all different types of networking. Cisco has forecasted that by 2020, 26.3 billion devices will be networked [4].

### 5.1.1 Enablers of IoT

The intriguing question, that arises right now, is: Why is IoT receiving so much attention right now? In other words, what have enabled the adoption of IoT technology? There is not a single factor that allowed the spread of IoT technology, but instead there were multiple technology achievements that enabled the rise of IoT. One of the main enablers of IoT has to be the ubiquitous wireless networking, which allows the connectivity of any device anywhere all the time. The progress in wireless networking, including mobile networks and small area networking (i.e. WiFi), has provided cheap means of wirelessly connecting a huge number of devices to the Internet; hence creating the term "Internet of Things". Relative to networking, the cost of bandwidth has also declined. This allowed more types of applications to be adopted, including high quality videos, which can be used in remote monitoring of properties, as well as remote medical diagnosis. This progress in wireless networking has been accompanied by widespread use of smartphones. The spread of Smartphones has helped with increased feasibility of IoT, since smartphones can act as personal gateways to IoT, acting as remote control or a hub. Basically, almost every IoT aspect has an interfacing application, which can be installed on a smartphone.

Moving from wireless networking, two other important enablers are cheap sensors and cheap processing. The advances in sensing technologies have been a major player in the spread of IoT applications. Such progress has dropped the costs of sensor fabrication, allowing the deployment of huge number of sensors in different applications, since they became more affordable. Additionally, the miniaturization of sensors, along with their reduced power requirements, represent another factor that enabled the deployment of many sensors, especially in small areas without much burden on small batteries used in many mobile devices [5].

On the other hand, processing costs have been in the decline, dropping nearly 60× over the last decade, enabling smart devices to be affordable yet with enough processing powers to understand their surroundings and automatically know what they should do without human control. Finally, big data and IPv6 were two more factors that enabled the addition of more devices to the connected world, forming part of the Internet of Things world. Obviously, IoT generates huge amounts of unstructured data, although may not require high data rates. The availability of big data allowed the high data volumes required by IoT. The move from IPv4 to IPv6 increased the number of possibly connected devices. IPv4, which supports only 32-bit addresses, allows the connectivity of approximately 4.3 billion devices, while moving to IPv6, using 128-bit addresses, increased the number of connected devices to approximately $3.4 \times 10^{38}$ devices [3].

## 5.1.2   Value Proposition of IoT

Many companies are already embracing the concept of IoT. IoT is used within those companies on two different premises. IoT technology can be used as a revenue generation itself, or can be used for increasing productivity and reducing costs. For revenue generation, companies focus on driving incremental reviews through the introduction of new IoT-based products and services. As per example, AT&T®, in partnership with car manufacturers such as Audi®, GM®, Tesla® and Volvo®, has introduced a Connected Car service. Such service offers a high speed connections, using 3G or 4G, on a monthly subscription basis (10$/month). Many opportunities exist for other services and products, in all areas of IoT, such as Smart City, ITS, eHealth, etc.

Alternatively, companies benefit from IoT applications, by reducing their costs and increasing productivity, through using smarter production or manufacture methods. IoT can be used to reduce costs of Capital costs (Capex), labor or energy. For instance, Verizon® is exploiting IoT to save energy, through the deployment of hundreds of sensors and control points throughout their data centers, connected wirelessly. As a result, Verizon® managed to decrease their energy consumption, up to 55 million kWh annually across 24 data centers, saving 66 million pounds of greenhouse gases per year [3].

### 5.1.3 Sensors in IoT

IoT systems, in general, consist of three domains: Sensors, Connectivity and Applications. Most attention has been focused on the application category. Connectivity has been well investigated as well; wireless networking is achieving and even surpassing their potential. In contrast, the underlying hardware of sensor technologies has been lacking enough efforts to keep up with the advances in IoT, specifically the links of those sensors to the 'cloud' [5].

Sensor technology is remarkably diverse. Sensors are usually very specialized and execute a certain basic very specific task, based on the application and performance requirements. For instance, in any settings, there would be many different devices, but all measuring temperature ones are based on one of the technologies, namely thermistors, bimetallic strips, and very occasionally thermocouples [6]. The fact, that there are too many devices doing the same sensing job, emphasizes two important basic points with regards to sensors and their implementation. The first point is that sensors are very specialized and are very limited to one certain application need. Second, the same technology, in an entirely different package, can fulfill a range of sometimes radically different requirements. It is important to understand that sensors are usually deployed in IoT, to enable or facilitate another task. In such environment and technology, how can one decide which sensor or technology to use? On the other hand, what can new technologies contribute to sensing techniques? Sensor technology experts have created three tests, of which at least one must be demonstrated by new technology. The first test is whether the new technology can perform the same job with the same specification but for tenth of the cost. The other test is for the technology to perform the same job but ten times better and for the same cost. Lastly, the question, to be asked, is if the technology can perform a measurement that has been hitherto impossible [6].

### 5.1.4 Fiber Sensors in IoT

The benefits of fiber sensing have already been well demonstrated and were previously discussed in this book. However, for those benefits to be advantageous, they have to provide some gains to potential users, when compared to what is currently available. Obviously, those benefits depend on the applicable market sector, along with current and projected needs of such market. Among those are the necessary qualities of stability, repeatability, resolution, accuracy, and reliability, across a range of environmental conditions. Additionally, some cultural momentum comes into play, which can particularly be seen in more conservative industries, such as aerospace. In such conservative industries, the need for established codes of practice is usually required prior to adopting a particular technology breakthrough, in addition to the somewhat contradictory parallel need to adopt such technology in order to verify the established code of practice. Those examples elaborate the challenging puzzle of actually adopting new sensing technologies in the practical world [6].

Despite the challenges facing the adoption of new technologies, including fiber-based sensing, OFSs have shown certain qualities making them suitable for many applications of the world of IoT. OFSs have demonstrated their ability for multi-plexing and distributed sensing at low-costs, offering an attractive solution for measurements of physical parameters over long distances or large areas. The full distributed OFS network (Rayleigh, Raman, and Brillouin scattering) is considered a 3S system (smart materials, smart structure, and smart skill). Distributed OFSs can be embedded in power grids, railways, bridges, tunnels, roads, dams, oil and gas pipelines among other facilities, and can be integrated with wireless networking [7]. The benefits of OFSs have been discussed thoroughly in previous chapters. Most of those benefits are still relevant in IoT applications.

Now that IoT has been introduced, the rest of the chapter discusses different applications of OFSs within IoT different categories, building on the knowledge gained about fiber-based sensing, its characteristics, pros and cons, in addition to examples of how to deploy OFSs for different applications and conditions, presented in previous chapter.

## 5.2 Smart Composite Materials and Structures

One of the applications, to start with, is composite materials and structures. The use of Composite materials varies in a wide range of fields, including sport/leisure, aerospace, maritime structures, transport, and civil engineering industries. Composite material structures are usually subjected to external perturbations and often harsh changing environmental conditions, which cause fatigue damage and/or failures to those structures; hence raising the need for real-time structural health monitoring (SHM) [8–11]. There is a great demand for real-time diagnosis process and condition monitoring of composite structures, while in place during their working lifetime [9–11]. The continuous monitoring aims at detecting, identifying and locating the defects within the structure that may threaten their safety and/or performance.

A variety of traditional nondestructive evaluation (NDE) techniques already exist to perform such SHM, varying between ultrasonic inspection, acoustography, low-frequency methods, radiographic inspection, shearography, acousto-ultrasonic, and thermography [12]. Although those techniques are considered effective, most of these techniques are difficult to use during operational mode of the structures, due to their size and weight [8, 12].

Alternatively, OFSs are better suited options for real-time SHM, during operational mode, since they can be easily embedded into composite structures acting as a nervous system [8]. OFSs have proved their potential, when compared to traditional NDE techniques. Initially they provide the typical advantages of optical fiber-based sensors, being minimal weight, small size, high bandwidth, and immunity to electromagnetic interference. Additionally, OFSs offer unique capabilities in the field of composite materials. They are suitable for monitoring the manufacturing process of composite parts. They can be used for nondestructive testing, once fabrication is complete, enabling health monitoring and structural control [8]. OFSs have proved capable of monitoring multiple parameters including stress/strain, temperature, composite cure process, vibration, humidity, delamination, and cracks [12, 13]. Furthermore, due to progress in optoelectronic industries and telecommunication, optical fibers are steadily and increasingly becoming a cost effective solution.

Finally, for sensors to be used in composite structures, they need to have minimal effect on the characteristics and functionalities of the hosting materials. OFSs perfectly fulfill this requirement, since they are compatible with reinforcement fabrics and can be embedded within its structures without modifying their strength [8].

The fabrication process is very critical, since the produced OFSs have to meet all the requirements, from performing its monitoring task perfectly to keeping the characteristics of the hosting structure unchanged. Many fabrication processes exist. However, for this specific application with its critical requirements, the expertise-intensive hand layup and pre-preg layup methods [14] are mostly adopted. In the hand layup process, fiber-reinforced layers are arranged in a laminate; then the laminate is figured to produce the required part. The process stacks the reinforcement fibers or fabrics over each other, while applying the matrix between them. Afterwards, the resulting laminate is shaped as required [8].

Pre-pregs are known to be single laminates of *pre-impregnated* composite fibers with a matrix material such as epoxy resin. The pre-preg layup method stacks multiples composite pre-preg laminates over each other. The curing is then performed unaided or by applying heat and/or pressure. The shapes are molded either using vacuum-bag molding or an autoclave molding [12].

A similar process is used in embedding optical fibers in samples. First, optical fibers are placed on the composite layer, while applying some pre-strain, to produce fibers free of bends [8, 15]. It is worth mentioning that the exact position of the OFS is application specific and is determined by the locations where parameters are to be monitored.

Multiple monitoring applications exist in the field of composite materials, including but not limited to monitoring vibration, cure process, and cracking detection [9, 11]. In all referenced applications, the main parameters to monitor are strain or temperature or both parameters, simultaneously. In the area of composite materials, more than one type of OFSs have been reported for strain/temperature monitoring, including interferometric OFS [16], fiber Bragg grating (FBG) [13], distributed sensors (using various techniques such as Rayleigh scattering, Raman scattering, and Brillouin scattering) [7, 17]], and polarimetric sensors [18, 19].

## 5.3 Smart Wearable Devices

Smart wearable devices are sensor based gadgets that are worn by individuals to monitor different parameters, whether physical parameters of surrounding environment or biological parameters within the body of the individual for health monitoring. Within this field, sensors are considered the most diverse component and the key enabler for smart wearable devices. Along current known wearable gadgets such as smart watches, smart textiles are becoming a major player in the field of smart wearable devices. Smart textiles can be defined as textiles, which can interact with their surrounding environment, through sensing certain parameters and either reacting based on those measurements or relaying those measurements to other control units for decision [20]. Different types of sensors can be embedded within smart textiles, which are capable of monitoring various physical parameters. Smart textiles have been proposed to be used in multiple fields, including transportation, medicine, civil engineering, etc. Medicine applications are considered one of the main fields of application of smart textiles. They can be used for the continuous monitoring of physiological parameters of great importance, including respiratory activity, heartbeat or blood pressure. The main goal for developing smart textiles for medical reasons is to increase the mobility of patients or elder citizens, who require all-time continuous medical surveillance [20].

Currently, OFSs are gaining much advantage compared to traditional electrical and mechanical sensors for the monitoring of thermal and mechanical parameters in the field of smart textiles. As usual, OFSs owe their advantages to their small size and flexibility, their good metrological properties, and of course—in this context—their immunity to electromagnetic field. This last property specifically allows smart textiles using OFSs to be used in a magnetic resonance imaging environment, in contrast to standard electronic sensors, which are hazardous in such environment. For instance, OFS-based smart textiles can be used for monitoring biological parameters (e.g. respiratory and heartbeat) during a magnetic resonance procedure [20].

Fiber optic technology perfectly fits in the application of smart textiles, since they can provide both sensing and signal transmission, simultaneously. Polymer optical fibers (POFs) are specifically perfect for smart textile implementation, since they are cheap, lightweight, flexible, robust, and able to measure high strain values without damage. Another example is gait analysis using wearable sensors, which is an inexpensive, convenient, and efficient alternative to traditional electronic solutions. It is capable of providing useful information for multiple health-related applications. In clinical experimentation in rehabilitation and diagnosis of certain medical conditions and sport activities, gait analysis using wearable sensors have shown great performance and potential [21]. During the last decade, intensive research efforts have focused on the integration of smart textiles and optical fiber technology [20, 22].

As it is clear now, OFSs can be manufactured using multiple principles. However for this specific application of smart wearable devices or smart textiles, sensors based on the fiber Bragg grating technology or intensity-based OFSs have been mostly used [20].

Finally, one issue with smart textiles using OFSs, that needs to be addressed if such technology is to be widely adopted, is the need to improve their stability against light coupling, specifically in homecare long-term applications [20].

## 5.4   eHealth Smart Applications

Another important application of sensors, gaining lots of attention in the IoT world, is in the field of healthcare. eHealth is a very rich environment, where sensors can be used for various applications, to facilitate management in hospitals or clinics, in addition to provide better quality healthy life to individuals in general, and patients in particular. Using sensors attached to the body of an individual, vital signs and physiological parameters can be continuously monitored; hence providing a ubiquitous (any time, everywhere) medical surveillance to patients. Some possible parameters that can be monitored are electrocardiography (ECG), respiration, skin conductance, and skin temperature. In the field of eHealth, the sensors can either be invasive or non-invasive, with different requirements. Invasive means the sensors are inserted inside the body of the patient/individual; hence they have to be miniaturized and biocompatible. On the other hand, non-invasive sensors are placed on the skin surface or very close to the body (as in the case of smart textiles) [20, 23].

OFSs are good option for use in the emerging field of eHealth, whether in invasive or non-invasive manner. They owe their advantages to their desirable metrological properties such as low zero drift, sensitivity drift, and good sensitivity. Clearly, OFSs are immune to electromagnetic interferences. All the above mentioned properties enabled OFSs to match other more traditional electrical and mechanical sensors, while surpassing them in many fields of application. Finally, the fact, that OFSs can be used in a Magnetic Resonance (MR)-compatible setup, further increases the research interest in optical-fiber based eHealth technologies [20]. Recent work has demonstrated body-monitoring systems, based on optical fibers and their application in the field of physiological monitoring [22, 26]. Despite all the advantages, some challenges are still facing the development of optical fiber based eHealth devices. There is an obvious need for the development of easy-to-use 'plug and play' devices, which are tolerant to rough handling. Additionally, the costs of necessary instrumentation still need to be reduced, if such technology is to reach its potential and be affordable by the majority of patients. Basically, application-oriented customizations, reduced costs and robust designs are critical factors for their commercialization, especially for the medical market [23].

## 5.5   Smart Cities

We live in a world, where resources are limited and scarce. Usually, urban areas consume the vast majority of these resources. For our survival on this planet, cities need to be more sustainable. Advanced automated systems to enhance processes within a city are foreseen to play a major role in the cities of the future. Some examples of those systems include smart buildings, which are capable of capturing

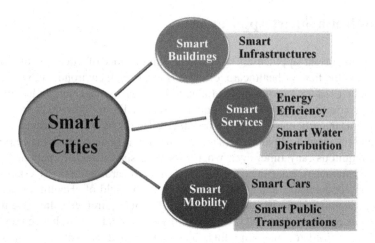

**Fig. 5.2** Different components of the "Smart City" concept

rain water for later use; and intelligent control systems to autonomously monitor infrastructures [27]. The potential possibilities are numerous. All those advanced automated systems lie under the umbrella of the "Smart City" concept.

A smart city can be seen as a city, whose infrastructures and services are cohesively integrated, while using intelligent devices for monitoring and control, to ensure sustainability and efficiency. The motivation behind intelligent innovative solutions to provide sustainable cities is clear. Population are constantly growing, significantly increasing resources consumption, causing shortage in resource and climate change. It has been clear that urban areas are particularly responsible for the major consumption; hence the increasing need for smarter infrastructures [27]. The "Smart City" solution comprises multiple smart components of urban dynamics, as illustrated in Fig. 5.2.

It is worth noting that clearly sensing plays a major role in the concept of "Smart City". Smart sensing provides a tool to automatically monitor certain parameters, then intelligently react on its own based on measured parameters. For one instance, smart sensors can be used to monitor large infrastructures, such as bridges, roads and buildings; hence providing a real-time continuous surveillance of those infrastructures. Such continuous monitoring would eliminate the need for regular scheduled inspections, therefore reducing costs. Another application is measuring energy consumption in households, allowing load forecasting [27].

As clear from the above arguments, sensors are at the heart of the "Smart City" application. There is a need for numerous sensors to be deployed and well connected. The collected data have to be intelligently acquired, analyzed and well processed, to make intellectual decisions [27]. The need for a large number of sensors raises a challenge for the production of such amount of sensors. Besides being small, accurate and resilient, the production of sensors has to be cost-effective. Meeting those requirements, OFSs represent a perfect fit for use in Smart City applications. In the rest of subsection, different usages of OFSs in the different components of Smart Cities are discussed.

### 5.5.1   Smart Buildings

In a Smart City setup, all standing infrastructures should be continuously monitored for safety and disaster avoidance. Although infrastructures in cities are usually periodically examined by experts through scheduled visits, a visual inspection is often not perfect and anomalies may sometime go undetected. Thanks to automation, a more efficient way exists through installing multiple sensors across the overall structure, which automatically monitor the conditions of buildings [11]. In the context of SHM, different types of sensors have been used, including corrosion rate sensors (using the principle of increased electrical resistivity due to corrosion); acoustic emission sensors (based on the propagation of sound waves); and magnetostrictive sensors (which are built on the concept of detecting changes in the magnetic induction of the material caused by strain or stress). Some more sophisticated sensors can even be used such as accelerometers, which measure both acceleration and vibration [11, 31].

Recently, OFSs have been used in the context of SHM [11]. For example, OFSs can be used in bridges and buildings for damage detection. As explained in previous chapter, OFSs, in this scenario, measure strain, pressure or temperature, by detecting the changes in a transmitted signal, after passing through the structure, by interpreting wavelength change. The OFSs need to be embedded within the structure to be monitored. The used OFSs can also be built based on distributed sensors or FBG sensors [28–31]. OFSs have already shown very promising performance in field tests.

### 5.5.2   Smart Services

"Smart services" is another field of the "Smart City" concept. This section lists two examples, where sensing is of critical importance. Those examples are monitoring of energy consumption and monitoring of water consumption/leakage. Those two services can contribute towards the sustainability of cities, through decreasing the consumption of energy and water, which are considered two of the most critical and scarce resources of our society.

Power consumption measurement can provide critical information, which can be used to decrease the overall energy consumption of household; and consequently that of the city as a whole, contributing towards the goal of sustainability of Smart City. Smart meters concept has been introduced using bi-directional grids [27]. Smart meters can help customers decrease their energy consumption; hence lowering their electricity bills, through usage of certain appliances during off-peak times, when energy is at its lowest price. Crucial to this concept is the precise monitoring of electronic appliances and measuring their energy consumption. Accurate measurement of power usage requires high sensitive sensors. Towards this end, sensors have been used to measure power consumption and monitor electrical appliances [27].

Besides energy (i.e. electricity), water would be considered the most vital resource in urban areas, which is already scarce. For that reason, water distribution has to be performed in the most efficient way possible. Additionally, quality control of water is one trait that cannot be compromised for the safety of the consuming citizens [32]. Obviously, continuous monitoring of water control is a vital requirement in urban environment. Some chemical sensing devices would be perfectly suitable for monitoring the quality of water, especially distributed for residential areas [11, 27].

The water distribution systems are basically non-intelligent. If a pipe gets damaged and water leakage occurs, it may be tricky and time consuming to detect such failure. It may even goes unnoticed for long time (may be years), in case of damage in underground pipes. Advanced sensing would provide a more reliable solution that can offer a fast fault detection system. Sensors are hence perfect for deploying in water distribution systems. They can be used to monitor water level in reservoir tanks, detect faults and leakage, and observe the water quality along the distribution system [33–35]. Based on the characteristics and sensitivity of OFSs, they represent a cost-effective efficient option for usage in water distribution systems, varying from monitoring water levels in reservoirs to detecting damage in pipes; hence consequently leakage.

### 5.5.3   Smart Mobility

With the steep increase in number of cars, there is an increasing need for efficient traffic management, to avoid traffic jams and optimize traffic flow, which would lead to more efficient driving, consequently saving energy contributing to the sustainability of Smart City.

Cars are evolving into smarter version, reaching optimally autonomous driving or the so-called "self-driving cars". Those foreseen smart cars should be able to sense other cars, know about traffic and make smart decisions based on the obtained information. The move towards smarter transportation is motivated by higher safety and more efficiency. To achieve such goals, different sensing capabilities are required. Self-driving cars need to sense other cars, as well as pedestrians and any obstacle on the road. To provide efficient transportation, traffic needs to be monitored, so smart cars would be able to choose faster routes based on traffic. Additionally, for safety of cars and their passengers, certain sensors are required within the cars themselves, such as temperature monitoring, fluid measurement, brake testing, etc. All those functionalities require a huge numbers of sensors to be implemented in our future smart mobility scenario. Those sensors have to be highly accurate and reliable. Different types of sensors have been already deployed in the first generation of smart mobility, and intelligent transportation system (ITS), including traditional electrical and mechanical sensors. This field has a potential market for OFSs, as well. OFSs can provide many required measurement capabilities in the field. For instance, pressure monitoring OFSs can be used in streets to monitor the amount of traffic in different roads. Moreover, temperature monitoring OFSs can be used in vehicles either to control the temperature of the vehicle to keep it at the comfort temperature for passengers, or monitoring the temperature of the engine, to alert if it gets too heated. This field of smart mobility and intelligent transport system (ITS) is rapidly evolving, and it will definitely have multiple opportunities for the use of OFSs.

## 5.6  Summary

This chapter has provided a quick look into the world of "Internet of Things". The chapter started with an introduction to IoT, its main enablers, motivation and components. The chapter then provided a vision of how IoT can be split into different categories.

The chapter then moved to the more relative topic of sensors in IoT. The chapter discussed the required characteristics of sensors to be deployed in IoT. Afterwards, some examples of IoT applications have been introduced. Different categories of IoT, where authors see that OFSs would fit and offer advantages over traditional electronic or mechanical sensors, are then discussed. The categories discussed were smart composite materials, smart wearable devices, eHealth and Smart City. Within each category, different applications have been presented, specifically those where OFSs have high potential.

## References

1. Al-Fuqaha, Ala, et al. "Internet of things: A survey on enabling technologies, protocols, and applications," *IEEE Communications Surveys & Tutorials* 17.4 (2015): 2347-2376.
2. Atzori, Luigi, Antonio Iera, and Giacomo Morabito. "The internet of things: A survey." *Computer networks* 54.15 (2010): 2787-2805.
3. Jankowski, Simona, et al. "The Internet of Things: Making sense of the next mega-trend." Goldman Sachs, 2014.
4. CISCO White paper, "CISCO Visual Network Index: Forecast and Methodology, 2015-2020", June 2016.
5. http://www.ppc-online.com/blog/fiber-deployments-and-the-internet-of-things
6. Culshaw, Brian. "Future Perspectives for Fiber-Optic Sensing." Optical Fiber Sensors: Advanced Techniques and Applications. Ed. Ginu Rajan. CRC Press, (2015). 521-544.
7. Zhang, Zaixuan, et al. "Recent progress in distributed optical fiber Raman photon sensors at China Jiliang University." Photonic Sensors 2.2 (2012): 127-147.
8. Ramakrishnan, Manjusha, Yuliya Semenova, and Ginu Rajan. "Optical Fiber Sensors for smart composite materials structures." (2015): 491.
9. Ma, Jianjun, and Anand Asundi. "Structural health monitoring using a fiber optic polarimetric sensor and a fiber optic curvature sensor-static and dynamic test." Smart materials and structures 10.2 (2001): 181.
10. Valinejadshoubi, Mojtaba, Ashutosh Bagchi, and Osama Moselhi. "Structural Health Monitoring of Buildings and Infrastructure." Structural Health Monitoring 1 (2016): 50371.
11. Ye, X. W., Y. H. Su, and J. P. Han. "Structural health monitoring of civil infrastructure using optical fiber sensing technology: A comprehensive review." The Scientific World Journal 2014 (2014).

12. Méndez, Alexis, and A. Csipkes. "Overview of fiber optic sensors for NDT applications." Nondestructive Testing of Materials and Structures. Springer Netherlands, 2013. 179-184.
13. Kinet, Damien, et al. "Fiber Bragg grating sensors toward structural health monitoring in composite materials: Challenges and solutions." Sensors 14.4 (2014): 7394-7419.
14. Ramakrishnan, Manjusha, et al. "Measurement of thermal elongation induced strain of a composite material using a polarization maintaining photonic crystal fiber sensor." *Sensors and Actuators A: Physical* 190 (2013): 44-51.
15. Ramakrishnan, Manjusha, et al. "The influence of thermal expansion of a composite material on embedded polarimetric sensors." *Smart Materials and Structures* 20.12 (2011): 125002.
16. Leng, J. S., and A. Asundi. "Real-time cure monitoring of smart composite materials using extrinsic Fabry-Perot interferometer and fiber Bragg grating sensors." Smart materials and structures 11.2 (2002): 249.
17. Zhou, G., and L. M. Sim. "Damage detection and assessment in fibre-reinforced composite structures with embedded fibre optic sensors-review."Smart Materials and Structures 11.6 (2002): 925.
18. Bieda, Marcin S., et al. "Polarimetric and Fiber Bragg Grating reflective hybrid sensor for simultaneous measurement of strain and temperature in composite material." *SPIE Optics+ Optoelectronics*. International Society for Optics and Photonics, 2015.
19. Domański, A. W., et al. "Polarimetric Optical Fiber Sensors for Dynamic Strain Measurement in Composite Materials." Acta Physica Polonica, A.124.3 (2013).
20. Massaroni, Carlo, Paola Saccomandi, and Emiliano Schena. "Medical smart textiles based on fiber optic technology: An overview." Journal of functional biomaterials 6.2 (2015): 204-221.
21. Tao, Weijun, et al. "Gait analysis using wearable sensors." Sensors 12.2 (2012): 2255-2283.
22. Quandt, Brit M., et al. "Body-Monitoring and Health Supervision by Means of Optical Fiber-Based Sensing Systems in Medical Textiles." Advanced healthcare materials 4.3 (2015): 330-355.
23. Silvestri, Sergio, and Emiliano Schena. *Optical-fiber measurement systems for medical applications*. INTECH Open Access Publisher, 2011.
24. Leitão, Cátia Sofia Jorge, et al. "Plastic optical fiber sensor for noninvasive arterial pulse waveform monitoring." IEEE Sensors Journal 15.1 (2015): 14-18.
25. Leitão, Cátia, et al. "Central arterial pulse waveform acquisition with a portable pen-like optical fiber sensor." Blood pressure monitoring 20.1 (2015): 43-46.
26. Leitão, C., et al. "Optical fiber sensors for central arterial pressure monitoring." Optical and Quantum Electronics 48.3 (2016): 1-9.
27. Hancke, Gerhard P., and Gerhard P. Hancke Jr. "The role of advanced sensing in smart cities." Sensors 13.1 (2012): 393-425.
28. Mesquita, Esequiel, et al. "Global overview on advances in structural health monitoring platforms." Journal of Civil Structural Health Monitoring 6.3 (2016): 461-475.
29. Costa Antunes, Paulo, et al. "Dynamic structural health monitoring of a civil engineering structure with a POF accelerometer." Sensor Review 34.1 (2014): 36-41.
30. Antunes, Paulo, et al. "Optical fiber sensors for static and dynamic health monitoring of civil engineering infrastructures: Abode wall case study."Measurement 45.7 (2012): 1695-1705.
31. Antunes, Paulo, et al. "Structural health monitoring of different geometry structures with optical fiber sensors." Photonic Sensors 2.4 (2012): 357-365.
32. Bilro, L., et al. "Design and performance assessment of a plastic optical fibre-based sensor for measuring water turbidity." *Measurement Science and Technology* 21.10 (2010): 107001.
33. Antunes, Paulo, et al. "Liquid level gauge based in plastic optical fiber."Measurement 66 (2015): 238-243.
34. da Costa Antunes, Paulo Fernando, et al. "Elevated water reservoir monitoring using optical fiber accelerometer." Instrumentation Science & Technology 41.2 (2013): 125-134.
35. Maria de Fátima, F. Domingues, et al. "Liquid hydrostatic pressure optical sensor based on micro-cavity produced by the catastrophic fuse effect."IEEE Sensors Journal 15.10 (2015): 5654-5658.

# Chapter 6
# Conclusion

The continuous progress in electronics has led to disruptive innovations in the field of networking, towards the formation of a ubiquitous wireless network, which paved the way to the world of Internet of Things (IoT). The Internet of Things stands for a fully connected world, providing independent automated intercommunication between smart "Things", which act accordingly, based on information they receive or perceive from surrounding environment. The term "Things" here refers to smart devices that exist in our world, being either a vehicle (smart car, bus, taxi, or train), or a device in the house/building/city (heating systems, water distribution systems, even a fridge or a coffee machine) or in health care system (i.e. eHealth, in smart monitoring of patients, or gadgets such as smart watch or fitbits), just to mention some examples.

IoT was made possible through multiple enabling technologies, including ubiquitous wireless networking, widespread of smartphones, higher speed of data rates, and big data technologies, among others. Moreover, the heart of the IoT implementation lies in the sensing technology, which basically enables all "Things" to acquire valuable information from the environment and other smart devices, in order for an informative decision or an action to be taken by one smart device or more.

In order for IoT to provide its utmost potential, the sensing technology implemented must provide highly accurate information. Therefore, there is a considerable amount of research being done in the field of sensing technologies, investigating multiple various techniques such as mechanical or electronic. However, in recent years, the field of optical fiber sensing has been rigorously researched. The high interest in this new promising technique has been verified by the consumption market growth, which is predicted to reach approximately US$4.5 Billion, by 2020 [1]. Such growth is a reflection of the advantages and outperformance offered by this type of sensing, when compared to electronic or mechanical technologies in the same category.

Optical Fiber Sensors (OFSs) are highly advantageous, since they are small in size, light weight, resilient to external harsh environment and immune to

© The Author(s) 2017
M.F.F. Domingues, A. Radwan, *Optical Fiber Sensors for IoT and Smart Devices*,
SpringerBriefs in Electrical and Computer Engineering,
DOI 10.1007/978-3-319-47349-9_6

electromagnetic interference, bio-compatible; rendering OFSs multifaceted solution for applications in the fields of structural health monitoring, eHealth, bio-chemical control analysis, energy power plants, smart city, etc.

Moreover, they can be produced using regular optical fibers typically used in telecommunication field, representing a cost effective solution with the added ability of data transmission along the same type of fiber.

Due to their increasing popularity, cost-effectiveness and high performance, numerous research efforts have been dedicated to study Optical Fiber Sensors, continuously working towards cheaper, more efficient and higher resilient sensors. This was the main motivation behind producing a brief covering the topic of Optical Fiber sensing, its multiple modulation techniques, its various applications, highlighting its advantages and limitations, and discussing proposed solutions to overcome such limitations, especially efforts towards reducing the production and implementation costs of OFSs.

The brief introduces the multiple modulations and demodulation techniques, along with their more common sensing applications. The modulation techniques vary between intensity-based, wavelength-based, frequency-based, and phase-based. One intriguing characteristic of OFSs is their ability to be implemented in a distributed manner, where an optical fiber can provide continuous measuring of certain parameters along its whole length. This specific property offers an appealing cost-effective solution. Following the same thoughts of cost-effectiveness, low-cost OFS production alternatives are introduced within the brief.

In addition to the typical silica optical fibers used in sensing, polymeric optical fiber (POF) paved its way into the optical fiber sensing market, as an economical alternative. This type of fiber can be produced considerably cheaper than silica optical fiber. Moreover, polymer based sensors, although with different sensitivities, can be as accurate as silica based optical fibers. More importantly, the optical sources and detectors required in the modulation for POF sensors are low-cost/cost effective.

Finally, IoT concept is rapidly spreading across every aspect in our world. CISCO has anticipated that by year 2020, around 26.3 billion devices will be interconnected to the network (cloud or Internet) [2]. This represents a great market opportunity. In the heart of this huge market, sensing is a key enabling technology, without which IoT cannot survive. OFSs provide cost-effective, high performance, resilient solution, that can be used in different applications of IoT, including Smart Housing, Intelligent Transportation System (ITS), structural health monitoring (SHM), and more critical in the field of smart healthcare (i.e. eHealth). OFSs offer great characteristics for implementation within biomedical and health monitoring devices, since they are small in size, light weight, and especially immune to electromagnetic interference. For instance, this last property enabled OFSs to be used in the critical environment that is Magnetic Resonance (MR)-compatible setup. The use of OFSs in IoT has been discussed, highlighting their offered advantages and their future potential uses.

In summary, this brief provides a complete overview of optical fiber sensing, for readers interested in the field, and especially for readers new to the field, looking for a perfect venue to use as a starting point of their research in such field of high potential future.

# References

1. Market Research Report, "Fiber Optic Sensors Global Market Forecast & Analysis 2015-2020", *ElectroniCast*, March 18, 2016
2. CISCO White paper, "CISCO Visual Network Index: Forecast and Methodology, 2015-2020", June 2016.

Printed in the United States
by Bookmasters

Printed in the United States
By Bookmasters